Collins

2027 GUIDE to the NIGHT SKY

Radmila Topalovic and Dominic Ford

Published by Collins
An imprint of HarperCollins Publishers
1 Robroyston Gate,
Glasgow G33 1JN
collins.reference@harpercollins.co.uk
collins.co.uk

HarperCollins Publishers
Macken House
39/40 Mayor Street Upper
Dublin 1
D01 C9W8
Ireland

In association with
Royal Museums Greenwich, the group name for the National Maritime Museum, Royal Observatory Greenwich, Queen's House and *Cutty Sark*
www.rmg.co.uk

First published 2026

The publishers wish to acknowledge their deep debt of gratitude to two of their late astronomy authors: Storm Dunlop FRAS, FRMetS (1942-2025) for his stewardship of the Guides to the Night Sky and Night Sky Almanac and the enormous contribution of his friend and co-author the late Wil Tirion (1943-2024) for his expert star charts and diagrams. They both leave a valuable and ongoing legacy.

A catalogue record for this book is available from the British Library

ISBN 978 0 00 881573 8

10 9 8 7 6 5 4 3 2 1

Printed in the United Kingdom by Pureprint

If you would like to comment on any aspect of this book, please contact us at the above address or online.
e-mail: collins.reference@harpercollins.co.uk

MIX
Paper | Supporting responsible forestry
FSC® C022913
www.fsc.org

Contents

Introduction

The aim of this Guide is to help people find their way around the night sky at any time of the year, by showing how the stars that are visible change from month to month, and by highlighting various events that occur during 2027. The objects and events described may be observed with the naked eye, or nothing more complicated than a pair of binoculars.

The conditions for observing naturally vary over the course of the year. During the summer, twilight may persist throughout the night and make it difficult to see the faintest stars. There are three recognized stages of twilight: civil twilight, when the Sun is less than 6° below the horizon; nautical twilight, when the Sun is between 6° and 12° below the horizon; and astronomical twilight, when the Sun is between 12° and 18° below the horizon. Full darkness occurs only when the Sun is more than 18° below the horizon. During nautical twilight, only the very brightest stars are visible. During astronomical twilight, the faintest stars visible to the naked eye may be seen directly overhead, but are lost at lower altitudes. As the diagram shows, full darkness never occurs from 24 May to 21 July at the latitude of London, and at Edinburgh nautical twilight persists throughout the whole night.

Moonlight will affect the visibility of objects, providing a natural form of light pollution. At Full Moon, it may be very difficult to see some of the fainter stars and objects, and even when the Moon is at a smaller phase it may seriously interfere with visibility if it is near the stars or planets in which you are interested. A full lunar calendar is given for each month and may be used to plan night sky observations. It may also be useful to look at moonrise and moonset times for your particular location.

The celestial sphere

All the objects in the sky (including the Sun, Moon and stars) appear to lie at some indeterminate distance on a large sphere, centred on the Earth. This ***celestial sphere*** has various reference points and features that are related to those of the Earth. If the Earth's rotational axis (an imaginary line through the North and South Poles and the core of the Earth) is extended, for example, it points to the North and South Celestial Poles. Similarly, the ***celestial equator*** lies in the same plane as the Earth's equator, it divides the sky into northern and southern hemispheres. This Guide is written for use in Britain and Ireland, therefore

The duration of twilight throughout the year at London and Edinburgh.

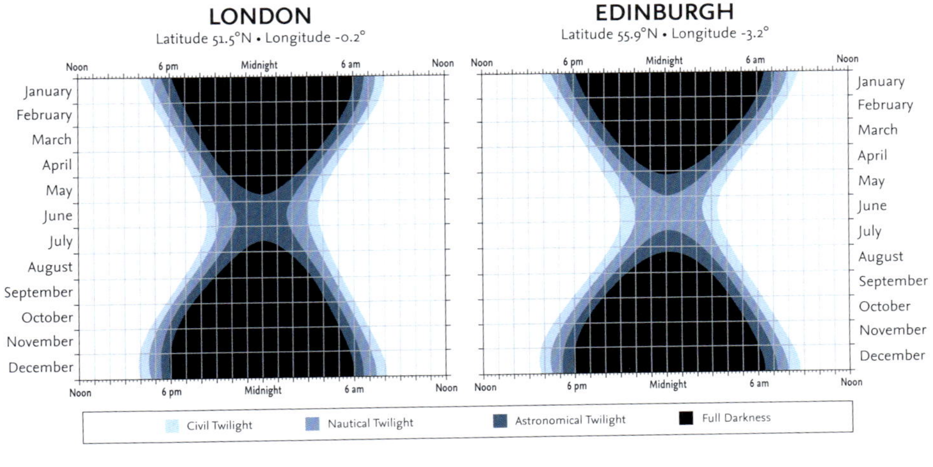

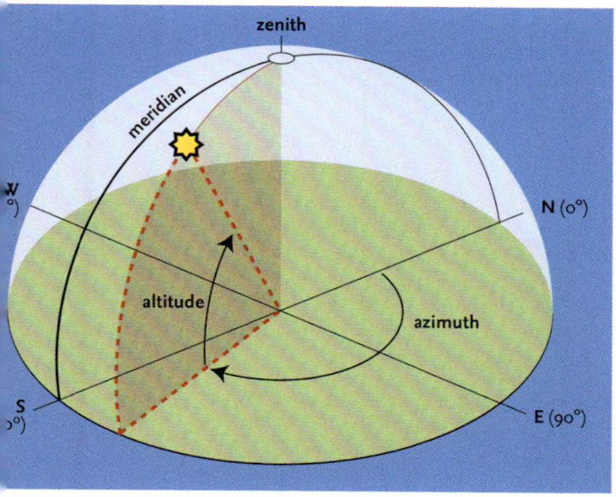

Measuring altitude and azimuth on the celestial sphere.

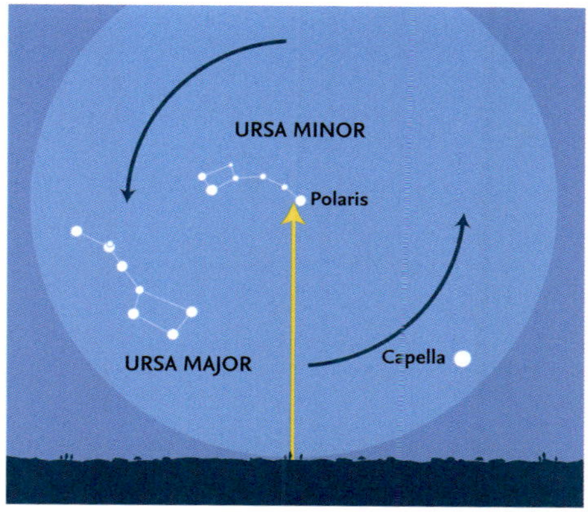

The altitude of the North Celestial Pole equals the observer's latitude.

the area of the sky that it describes includes the whole of the northern celestial hemisphere and those portions of the southern sky that become visible at different times of the year. Stars in the far south, however, remain below the horizon throughout the year, and are not included.

It is useful to know some of the astronomy terms for various parts of the sky. As seen by an observer, half of the celestial sphere is invisible at any point in time; these objects will be below the horizon. The point directly overhead is known as the ***zenith***, and the (invisible) one below one's feet as the ***nadir***. The line running from the north point on the horizon, up through the zenith and then down to the south point is the ***meridian***. This is an important invisible line in the sky, because objects are highest in the sky, and thus easiest to see, when they cross the meridian. Objects are said to ***transit*** when they cross this line in the sky.

In this book, reference is frequently made in the text and in the diagrams to the standard compass points around the horizon. The position of any object in the sky specific to the observer's location on Earth may be described by its ***altitude*** (measured in degrees above the horizon), and its ***azimuth*** (measured in degrees from north 0°, through east 90°, south 180° and west 270°). Experienced amateurs and professional astronomers also use another system of specifying locations on the celestial sphere called right ascension and declination, but here the simpler method will suffice.

The celestial sphere appears to rotate about an invisible axis, running between the North and South Celestial Poles. The location (i.e. the altitude) of the Celestial Poles depends entirely on the observer's latitude on Earth. At the North Pole (latitude 90°), the North Celestial Pole (NCP) would be directly overhead (at the zenith, or at an altitude of 90°). The charts in this book are produced for the latitude of 50°N, so the NCP is 50° above the northern horizon. The fact that the NCP is fixed relative to the horizon means that all the stars within 50° of the pole are always above the horizon and may, therefore, always be seen at night, regardless of the time of year. This northern circumpolar region is an ideal place to begin learning the sky, and ways to identify the circumpolar stars and constellations will be described shortly.

The ecliptic and the zodiac

Another important line on the celestial sphere is the Sun's apparent path against the background stars – in reality the result of the Earth's orbit around the Sun. This is known as the ***ecliptic***. The point where the Sun, apparently moving along the ecliptic, crosses the celestial equator from south to north is known as the vernal (or spring) equinox, which occurs on

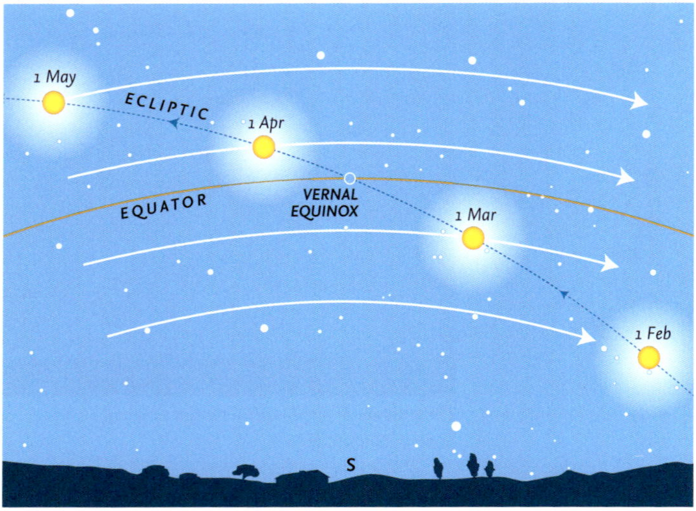

The Sun crossing the celestial equator in spring.

20 March. At this time (and at the autumnal equinox, on 23 September 2027, when the Sun crosses the celestial equator from north to south) day and night are almost exactly equal in length. The vernal equinox is currently located in the constellation of Pisces; this means the Sun is moving through Pisces on this day. It is important in astronomy because it defines the zero point for the system of celestial coordinates (right ascension) not used in this Guide.

The Moon and planets are to be found in a band of sky all the way around the celestial sphere that extends 8° on either side of the ecliptic. This is because the orbits of the Moon and planets are inclined at various angles to the ecliptic (i.e. to the plane of the Earth's orbit). This band of sky is known as the zodiac and, when originally devised, consisted of twelve ***constellations***, all of which were considered to be exactly 30° wide. When the constellation boundaries were formally established by the International Astronomical Union in 1930, adjustments were made resulting in the ecliptic passing through thirteen constellations and the Moon and planets passing through several other constellations that are adjacent to the original twelve.

The constellations

Since ancient times, the celestial sphere has been divided into various constellations, most dating back to antiquity and usually associated with certain myths or legendary people and animals. The boundaries of the 88 constellations across the whole celestial sphere have been fixed by international agreement and their names (in Latin) are largely derived from Greek or Roman originals. Some of the names of the most prominent stars are of Greek or Roman origin, but many are derived from Arabic names. Many bright stars have no individual names and, for many years, stars were identified by terms such as 'the star in Hercules' right foot'. A more sensible scheme was introduced by the German astronomer Johannes Bayer in the early seventeenth century. Following his scheme – which is still used today – most of the brightest stars are identified by a Greek letter followed by the genitive form of the constellation's Latin name. An example is the Pole Star, also known as Polaris and α Ursae Minoris (abbreviated α UMi). The Greek alphabet is shown on page 109, along with a list of all the constellations that may be seen from latitude 50°N, together with abbreviations, their genitive forms and English names. Other naming schemes exist for fainter stars; these are not used in this book.

Asterisms

Apart from the constellations, certain groups of stars, which may form a part of a larger constellation or cross several constellations, are readily recognizable and have been given individual names. These groups are known as ***asterisms***, and the most famous (and well-known) is 'the Plough', the common name for the seven brightest stars in the constellation of ***Ursa Major***, the Great Bear. The names and details of some asterisms mentioned in this book are given in the list on page 110.

Magnitudes

The brightness of a star, planet or other body is given in magnitudes (mag.). This is a mathematically defined scale where larger (positive) numbers indicate a fainter object. The scale extends beyond the zero point to negative numbers for very bright objects. (Sirius, the brightest star in the night sky is mag. -1.4.) Most observers are able to see stars as faint as about mag. 6, under very clear and dark skies.

The Moon

Although the daily spin of the Earth carries the sky from east to west (objects in the south rise in the east and set in the west) the Moon gradually moves eastwards relative to the background stars by approximately its diameter in an hour. This is equivalent to about half a degree across the sky. Its apparent eastward motion and change in phase are caused by its orbit around the Earth and changing position relative to the Sun.

Normally, in its orbit around the Earth, the Moon passes above or below the direct line between the Earth and the Sun (at New Moon) or outside the area obscured by the Earth's shadow (at Full Moon), explained in the diagram on page 17. Occasionally, however, the three bodies are more or less perfectly aligned to give an ***eclipse***: a solar eclipse at New Moon or a lunar eclipse at Full Moon. Depending on the exact circumstances, a solar eclipse may be merely partial (when the Moon does not cover the whole of the Sun's disk); annular (when the Moon is too far from Earth in its orbit to appear large enough to hide the whole of the Sun); or total. Total and annular solar eclipses are visible from very restricted areas of the Earth, but partial solar eclipses are normally visible over a wider area.

Somewhat similarly, at a lunar eclipse, the Moon may pass through the lighter outer zone of the Earth's shadow, the ***penumbra***, called a penumbral eclipse, which is not generally perceptible to the naked eye; part of the Moon could pass within the darkest part of the Earth's shadow, the ***umbra***, in a partial eclipse; or it could move completely within the umbra in a total eclipse. Unlike solar eclipses, lunar eclipses are visible from large areas of the Earth.

Occasionally, as it moves across the sky, the Moon passes between the Earth and individual planets or distant stars, giving rise to an ***occultation***. As with solar eclipses, such occultations are visible from restricted areas of the world.

The planets

The planets are named from the ancient Greek πλανήτης (planḗtēs) meaning wanderer, as they are always moving against the background stars. They are therefore treated in some detail in the monthly pages and information is given regarding when they are close to the Sun, other planets, the Moon or any of five bright stars that lie near the ecliptic. Such events are known as ***appulses*** or, more frequently, as ***conjunctions***. (There are technical differences in the way these terms are defined – and should be used – in astronomy, but these need not concern us here.) The positions of the planets are shown for every month on a special chart of the ecliptic.

The conditions of most favourable visibility depend on whether the planet is one of the two known as ***inferior planets*** – planets that orbit the Sun within the Earth's orbit (Mercury and Venus) and are consistently visible to the naked eye in the right conditions – or one of the five superior planets – planets that orbit the Sun beyond the Earth. Of the latter, only three (Mars, Jupiter and Saturn) are covered in detail, as they are visible to the naked eye. Occasionally, details of the fainter superior planets, Uranus

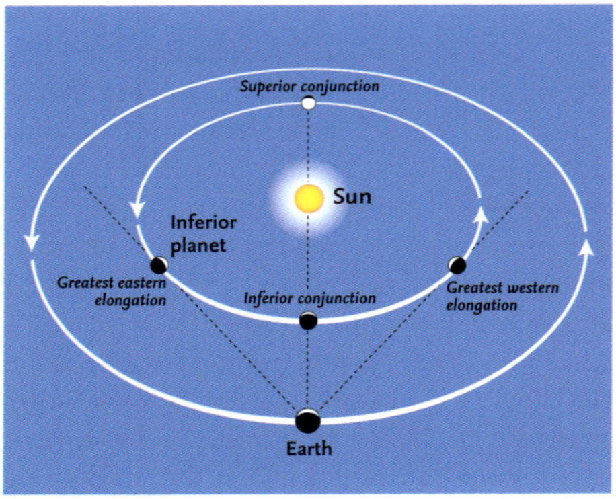

Inferior planet.

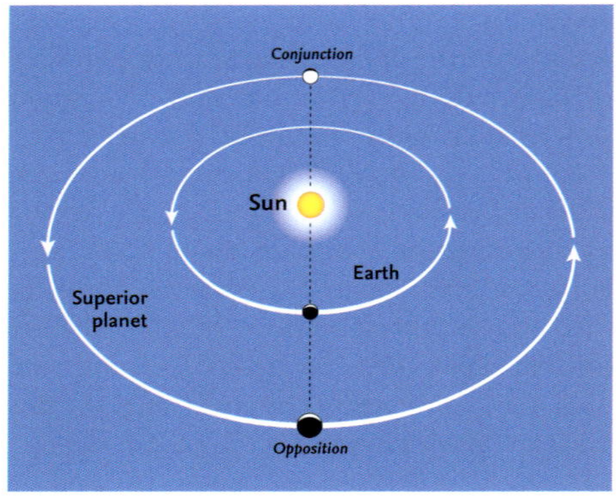

Superior planet.

and Neptune, are included, and special charts are given for them on page 25.

The inferior planets are most readily seen at eastern or western elongation, when their angular distance from the Sun (separation from the Sun in degrees) is greatest. Superior planets are best seen at ***opposition***, when they are directly opposite the Sun in the sky and cross the meridian at local midnight.

Angular distance can be measured approximately by holding one hand at arm's length. The various angles are shown in the diagram, together with the separations of the various stars in the Plough.

Meteors

At some time or other, nearly everyone has seen a meteor – a 'shooting star' – as it flashed across the sky. The particles that cause meteors – known technically as 'meteoroids' – normally range in size from that of a grain of sand (or even smaller) to the size of a pea. ***Fireballs*** or ***bolides*** are very bright meteors (brighter than magnitude -4) that are caused by objects up to 1 metre in size. Fireballs sometimes cause sonic booms that may be heard some time after the meteor is seen. On any night of the year there are occasional meteors, known as sporadics, that may travel in any direction. These occur at a rate that is normally between three and eight per hour.

Far more important, however, are meteor showers, which occur at fixed periods of the year, when the Earth encounters a trail of particles left behind by a comet or, very occasionally, by a minor planet (asteroid). Meteors of this type always appear to diverge from a single point on the sky, known as the radiant, and the radiants of major showers are shown on the charts. Meteors that come from a circular area 8° in diameter around the radiant are classed as belonging to the particular shower. All others that do not come from that area are sporadics (or, occasionally from another shower that is active at the same time). A list of the major meteor showers is given on page 31. Examples of meteors are shown on pages 32, 53, 77 and 101.

Looking directly at the radiant is not the most effective way of seeing meteors. They are most likely to be noticed if one is looking about 40–45° away from the radiant position. This is approximately two hand-spans as shown in the diagram for measuring angles.

Other objects

In the late eighteenth century Charles Messier, a French astronomer, compiled a catalogue of objects while he was searching for comets. These nebulae, clusters of stars and galaxies were given 'Messier numbers' (some already had names given several thousand years ago such as Praesepe – the Beehive Cluster). Some, such as the Andromeda Galaxy, M31, and the Orion Nebula, M42, may be seen by the naked eye, but all those given in the list

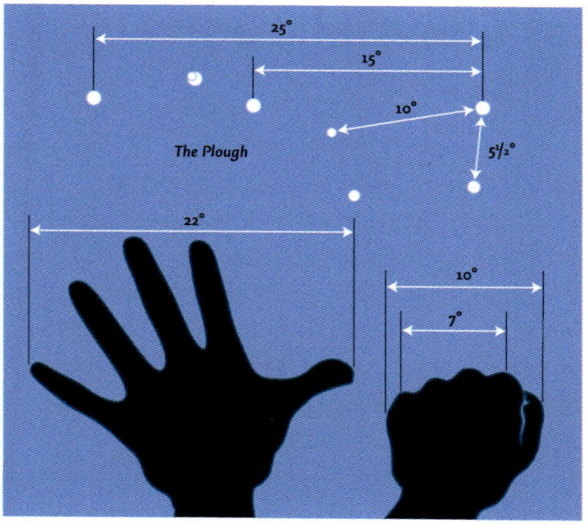

Measuring angles in the sky.

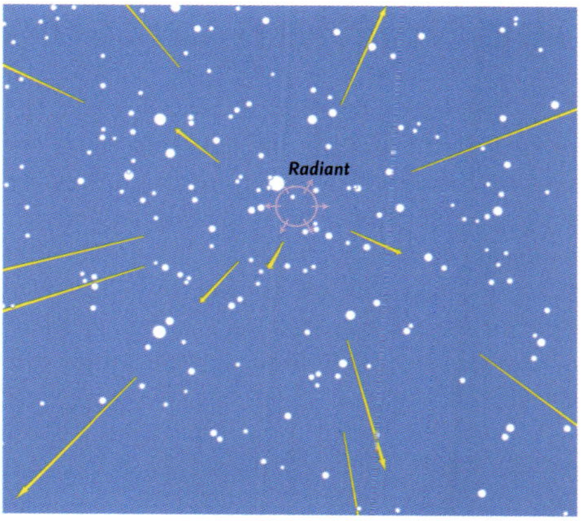

Meteor shower (showing the April Lyrid radiant).

will benefit from the use of binoculars. Apart from galaxies, such as M31, which contain thousands of millions of stars, there are also two types of cluster: open clusters, such as M45, the Pleiades, which may consist of a few dozen to some hundreds of stars; and globular clusters, such as M13 in Hercules, which are spherical concentrations of many thousands of stars. One or two gaseous nebulae, consisting of gas illuminated by stars within them, are also visible. The Orion Nebula, M42, is one, and is illuminated by a group of four stars, known as the Trapezium, which may be seen by using a good pair of binoculars.

Some interesting objects.

Messier / NGC	Name	Type	Constellation	Maps (months)
—	*Hyades*	open cluster	Taurus	Sep–Mar
—	*Double Cluster*	open cluster	Perseus	All year
—	*Melotte 111 (Coma Cluster)*	open cluster	Coma Berenices	Jan–Aug
M3	—	globular cluster	Canes Venatici	Feb–Aug
M8	*Lagoon Nebula*	gaseous nebula	Sagittarius	Jul–Aug
M11	*Wild Duck Cluster*	open cluster	Scutum	Jun–Oct
M13	*Hercules Cluster*	globular cluster	Hercules	Mar–Oct
M15	—	globular cluster	Pegasus	Jun–Dec
M20	*Trifid Nebula*	gaseous nebula	Sagittarius	Jul–Aug
M22	—	globular cluster	Sagittarius	Jul–Sep
M27	*Dumbbell Nebula*	planetary nebula	Vulpecula	May–Dec
M31	*Andromeda Galaxy*	galaxy	Andromeda	All year
M35	—	open cluster	Gemini	Oct–Apr
M42	*Orion Nebula*	gaseous nebula	Orion	Nov–Mar
M44	*Praesepe*	open cluster	Cancer	Nov–Jun
M45	*Pleiades*	open cluster	Taurus	Aug–Apr
M57	*Ring Nebula*	planetary nebula	Lyra	Apr–Dec
M67	—	open cluster	Cancer	Dec–May
NGC 752	—	open cluster	Andromeda	Aug–Mar
NGC 3242	*Ghost of Jupiter*	planetary nebula	Hydra	Mar–Apr

The Northern Circumpolar Constellations

The northern circumpolar stars are the key to starting to identify the constellations. For anyone in the northern hemisphere they are visible at any time of the year, and nearly everyone is familiar with the seven stars of the Plough – known as the Big Dipper in North America – an asterism that forms part of the large constellation of ***Ursa Major*** (the Great Bear).

Ursa Major

Like all the circumpolar constellations, due to the Earth's orbit around the Sun, Ursa Major lies in different parts of the evening sky at different periods of the year. The diagram below shows its position at the beginning of the four main seasons. The seven stars of the Plough remain visible throughout the year anywhere north of latitude 40°N. Even at the latitude (50°N) for which the charts in this book are drawn, many of the stars in the southern portion of the constellation of Ursa Major are hidden below the horizon for part of the year or (particularly in late summer) are too faint to be seen late in the night.

Polaris and Ursa Minor

The two stars ***Dubhe*** and ***Merak*** (α and β Ursae Majoris, respectively), farthest from the 'tail' are known as 'the Pointers'. A line from Merak to Dubhe, extended about five times their separation, leads to the Pole Star, ***Polaris***, or α Ursae Minoris. All the stars in the northern sky appear to rotate around it. There are five main stars in the constellation of ***Ursa Minor***, and the two farthest from the Pole, ***Kochab*** and ***Pherkad*** (β and γ Ursae Minoris, respectively), are known as 'the Guards'.

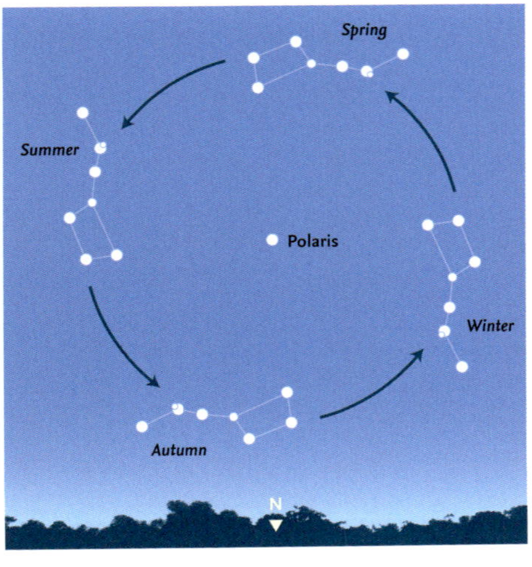

Cassiopeia

Facing Ursa Major on the other side of Polaris lies ***Cassiopeia***. It is highly distinctive, appearing as five stars forming a letter 'W' or 'M' depending on its orientation. Provided the sky is reasonably clear of clouds, you will nearly always be able to see either Ursa Major or Cassiopeia, and thus be able to orientate yourself on the sky.

To find Cassiopeia, start with ***Alioth*** (ε Ursae Majoris), the first star in the tail of the Great Bear. A line from this star extended through Polaris points directly towards γ Cassiopeiae, the central star of the five.

Cepheus

Although the constellation of ***Cepheus*** is fully circumpolar, it is not nearly as well-known as Ursa Major, Ursa Minor or Cassiopeia, partly because its stars are fainter. Its shape is rather like the gable end of a house. The line from the Pointers through Polaris, if extended, leads to ***Errai*** (γ Cephei) at the 'top' of the 'gable'. The brightest star, ***Alderamin*** (α Cephei) lies in the Milky Way region, at the 'bottom right-hand corner' of the figure.

Draco

The constellation of ***Draco*** consists of a quadrilateral of stars, known as 'the Head of Draco' (and also 'the Lozenge'), and a long chain of stars forming the neck and body of the dragon. To find the Head of Draco, locate the two stars ***Phecda*** and ***Megrez*** (γ and δ Ursae Majoris) in the Plough, opposite the Pointers.

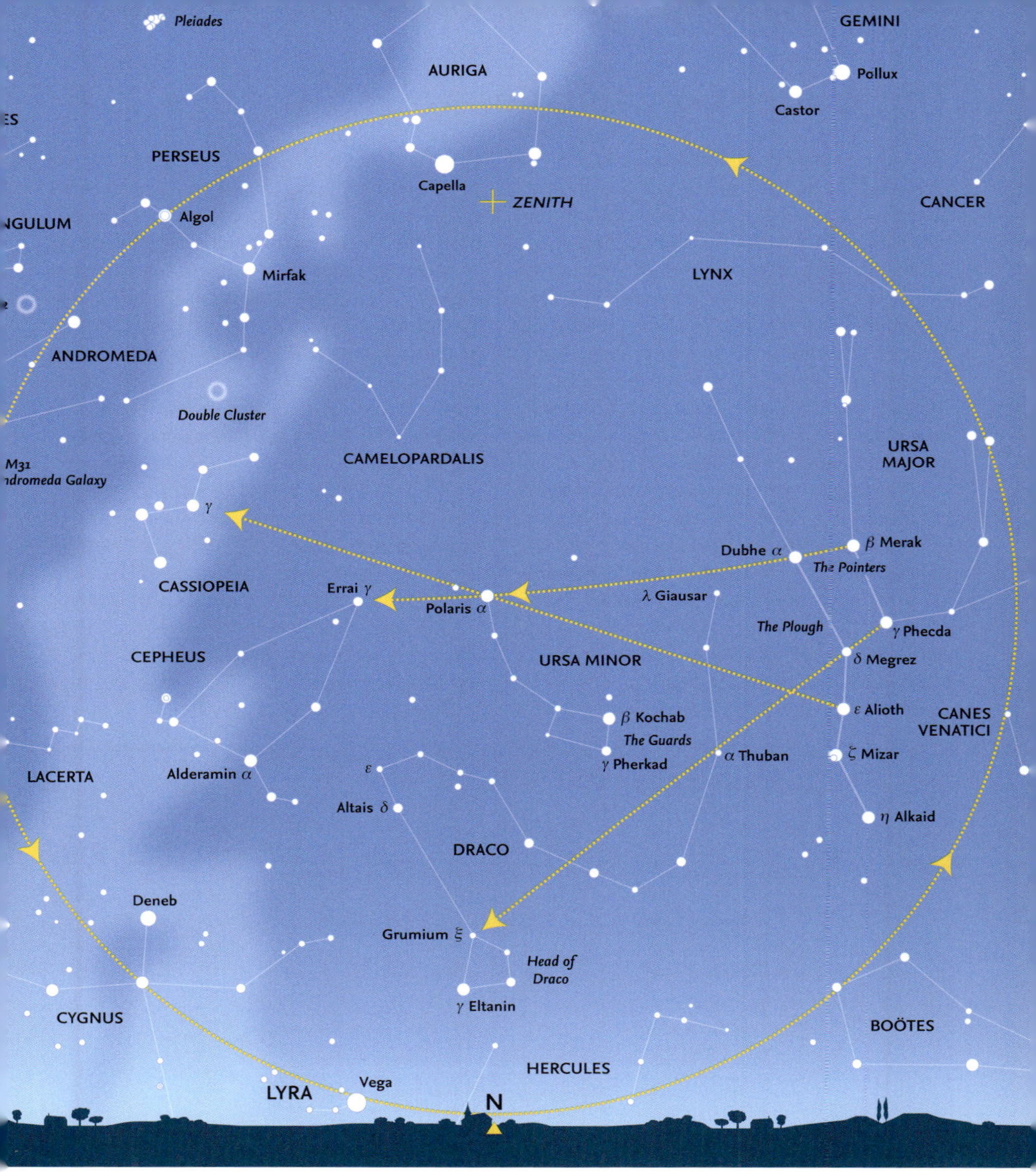

The stars and constellations inside the circle are always above the horizon, seen from latitude 50° N.

Extend a line from Phecda through Megrez by about eight times their separation, right across the sky below the Guards in Ursa Minor, ending at ***Grumium*** (ξ Draconis) at one corner of the quadrilateral. The brightest star, ***Eltanin*** (γ Draconis), lies farther to the south.

From the head of Draco, the constellation first runs northwest to ***Altais*** (δ Draconis) and ε Draconis, then doubles back southwards before winding its way through ***Thuban*** (α Draconis) and ending at ***Giausar*** (λ Draconis) almost between the Pointers and Polaris.

The Winter Constellations

The winter sky is dominated by several bright stars and distinctive constellations. The most conspicuous constellation is ***Orion*** (the Hunter), the main body of which has an hourglass shape. It straddles the celestial equator and is thus visible from anywhere in the world. The three stars that form 'the Belt' of Orion point down towards the southeast and to ***Sirius*** (α Canis Majoris), the brightest star in the night sky. ***Mintaka*** (δ Orionis), the star at the northwestern end of the Belt, farthest from Sirius, actually lies just slightly south of the celestial equator.

The bright red supergiant star ***Betelgeuse*** (α Orionis) lies above the belt, as does ***Bellatrix*** (γ Orionis). A line from Bellatrix at the 'top right-hand corner' of Orion, through ***Aldebaran*** (α Tauri), past the 'V' of the Hyades cluster, points to the distinctive cluster of bright blue stars known as the ***Pleiades***, or 'the Seven Sisters'. Aldebaran is one of the five bright stars that may sometimes be occulted (hidden) by the Moon.

On the far eastern side of Orion sits the constellation of ***Leo*** (the lion), a prominent constellation in the spring sky, easy to find with its backwards question mark (the 'head' of the lion).

Six bright stars in six different constellations: ***Capella*** (α Aurigae), Aldebaran (α Tauri), ***Rigel*** (β Orionis), Sirius (α Canis Majoris), ***Procyon*** (α Canis Minoris) and ***Pollux*** (β Geminorum) form what is sometimes known as 'the Winter Hexagon'. Pollux is accompanied to the northwest by the slightly fainter star of ***Castor*** (α Geminorum), the second 'Twin' of ***Gemini***.

In a counterpart to the famous asterism 'the Summer Triangle', an almost perfect equilateral triangle, 'the Winter Triangle', is formed by Betelgeuse (α Orionis), Sirius (α Canis Majoris) and Procyon (α Canis Minoris).

Several of the stars in this region of the sky show distinctive tints: Betelgeuse (α Orionis) is reddish, Aldebaran (α Tauri) is orange and Rigel (β Orionis) is blue-white.

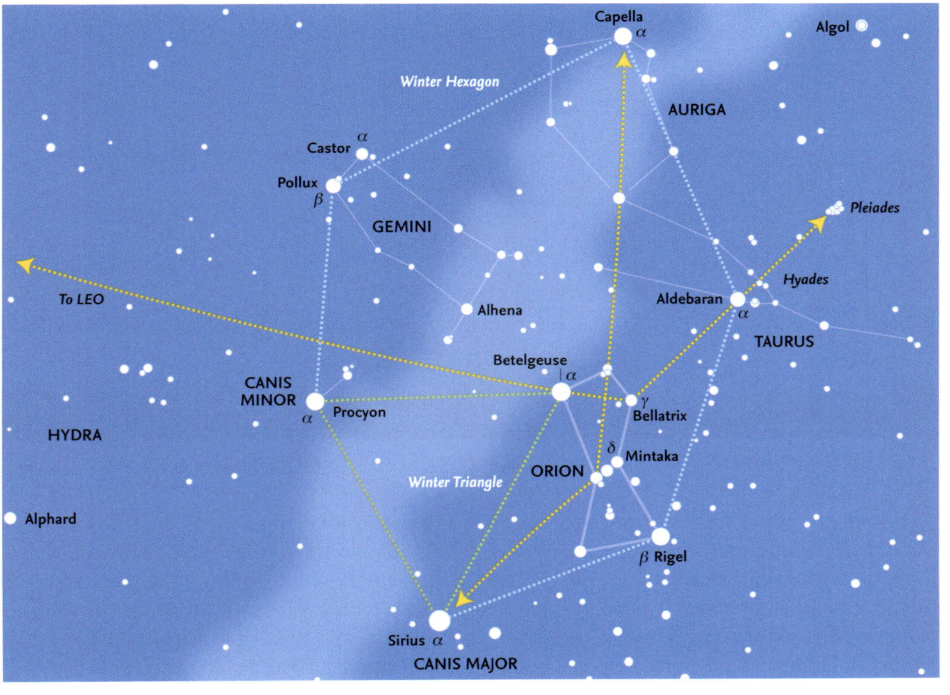

The Spring Constellations

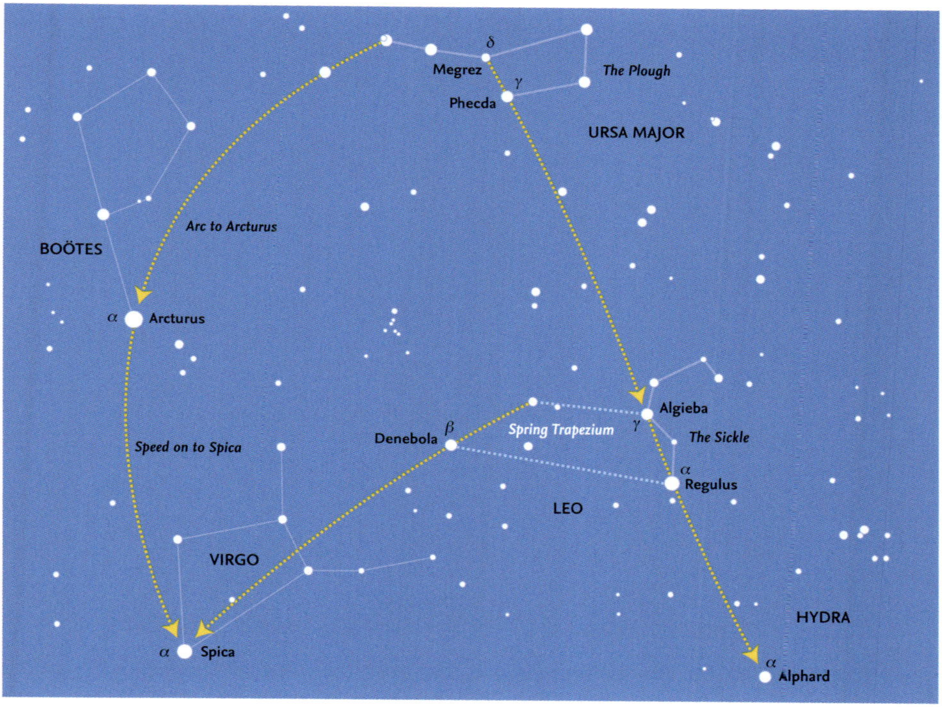

The most prominent constellation in the spring sky is the zodiacal constellation of ***Leo***, and its brightest star, ***Regulus*** (α Leonis), which may be found by extending a line from ***Megrez*** and ***Phecda*** (δ and γ Ursae Majoris, respectively) – the two stars on the opposite side of the bowl of the Plough from the Pointers – down to the southeast. Regulus forms the 'dot' of the backwards question mark known as 'the Sickle'. Regulus, like Aldebaran in ***Taurus***, is one of the bright stars that lie close to the ecliptic, and that are occasionally occulted by the Moon. The same line from ***Ursa Major*** to Regulus, if continued, leads to ***Alphard*** (α Hydrae), the brightest star in ***Hydra***, the largest of the 88 constellations.

The shape formed by the body of Leo is sometimes known as 'the Spring Trapezium'. At the other end of the constellation from Regulus is ***Denebola*** (β Leonis), and the line forming the back of the constellation through Denebola points to the bright star ***Spica*** (α Virginis) in the constel ation of ***Virgo***, a rough quadrilateral of moderately bright stars and fainter lines of stars extending outwards. A saying that helps to locate Spica is well-known to astronomers: 'Arc to Arcturus and then speed on to Spica.' This suggests following the arc of the tail of Ursa Major to ***Arcturus*** (α Boötis) and then a straight line on to Spica. Arcturus is actually the brightest star in the northern hemisphere of the sky. (Although other stars, such as Sirius, are brighter, they are all in the southern hemisphere.) Overall, the constellation of ***Boötes*** is sometimes described as 'kite-shaped' or 'shaped like the letter P'.

The Summer Constellations

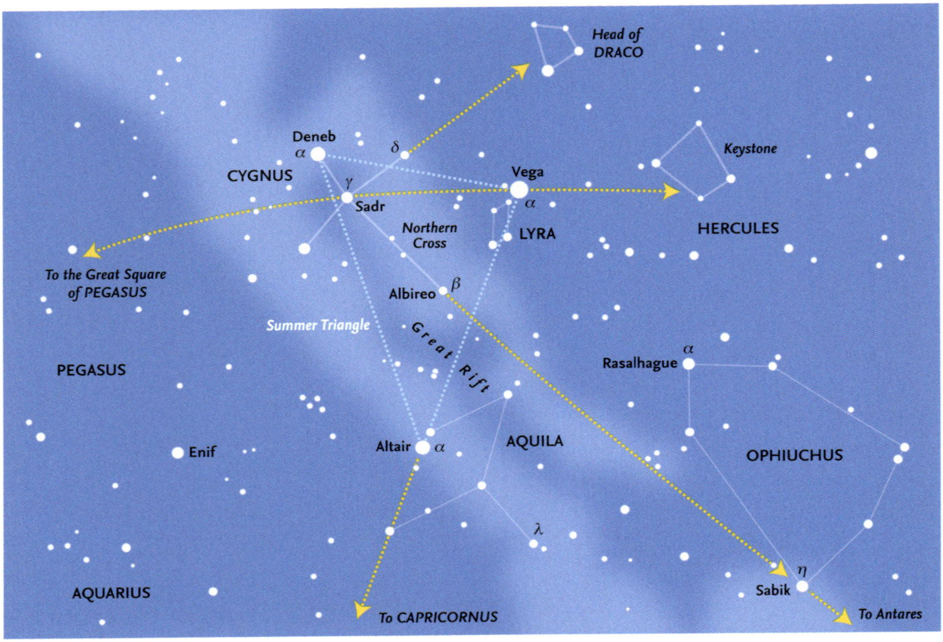

On summer nights, the three bright stars ***Deneb*** (α Cygni), ***Vega*** (α Lyrae) and ***Altair*** (α Aquilae) form the striking Summer Triangle (the asterism is still visible throughout most of Autumn). The constellations of ***Cygnus*** (the Swan) and ***Aquila*** (the Eagle) represent birds 'flying' down the length of the Milky Way. This part of the Milky Way contains the Great Dark Rift, an elongated dark region where the light from distant stars is obscured by intervening dust. The Dark Rift is clearly visible even to the naked eye in a dark sky region.

The most prominent stars of Cygnus are sometimes known as 'the Northern Cross' (as a counterpart to 'the Southern Cross' – the constellation of ***Crux*** – in the southern hemisphere). The central line of Cygnus through ***Albireo*** (β Cygni), extended well to the southwest, points to ***Sabik*** (η Ophiuchi) in the large, sprawling constellation of ***Ophiuchus*** (the Serpent Bearer) and beyond to ***Antares*** (α Scorpii) in the constellation of ***Scorpius.*** Like ***Cepheus***, the shape of Ophiuchus somewhat resembles the gable end of a house, and the brightest star ***Rasalhague*** (α Ophiuchi) is at the 'apex' of the 'gable'.

A line from ***Sadr*** (γ Cygni) to Vega indicates the central portion, 'the Keystone', of the constellation of ***Hercules. Draco*** is adjacent on the eastern side of Hercules. An arc through the same stars, in the opposite direction, points towards the constellation of ***Pegasus***, and more specifically to 'the Great Square of Pegasus'.

Aquila is less conspicuous than Cygnus and consists of a diamond shape of stars, representing the body and wings of the eagle, together with a rather faint star, λ Aquilae, marking the 'head'. ***Lyra*** (the Lyre) mainly consists of Vega (α Lyrae) and a small quadrilateral of stars to its southeast. Continuation of a line from Vega through Altair lands on the zodiacal constellation of ***Capricornus.***

The Autumn Constellations

During the autumn season, the most striking feature is the Great Square of Pegasus, an almost perfect rectangle on the sky, forming the main body of the constellation of ***Pegasus***. However, the star at the northeastern corner, ***Alpheratz***, is actually α Andromedae, and part of the adjacent constellation of ***Andromeda***. A line from ***Scheat*** (β Pegasi) at the northwestern corner of the Square, through ***Matar*** (η Pegasi), points in the general direction of ***Cygnus***. A line from ***Markab*** through the last star in the Square, ***Algenib*** (γ Pegasi) points in the general direction of the five stars, including ***Menkar*** (α Ceti), that form the 'tail' of the constellation of ***Cetus*** (the Whale). A ring of seven stars lying below the southern side of the Great Square is known as 'the Circlet', part of the constellation of ***Pisces*** (the Fish).

Extending the line of the western side of the Great Square towards the south leads to the isolated bright star, ***Fomalhaut*** (α Piscis Austrini), in the Southern Fish. Following the line of the eastern side of the Great Square towards the north leads to ***Cassiopeia*** while, in the other direction, it points towards ***Diphda*** (β Ceti), which is actually the brightest star in Cetus.

Three bright stars leading northeast from Alpheratz form the main body of the constellation of Andromeda. Continuation of that line leads towards the constellation of ***Perseus*** and ***Mirfak*** (α Persei). Running southwards from Mirfak is a chain of stars, one of which is the famous variable star ***Algol*** (β Persei), a star whose brightness changes periodically every 2.9 days. Farther east, an arc of stars leads to the prominent cluster of the ***Pleiades***, in the constellation of ***Taurus***.

Between Andromeda and Cetus lie the two small constellations of ***Triangulum*** (the Triangle) and ***Aries*** (the Ram).

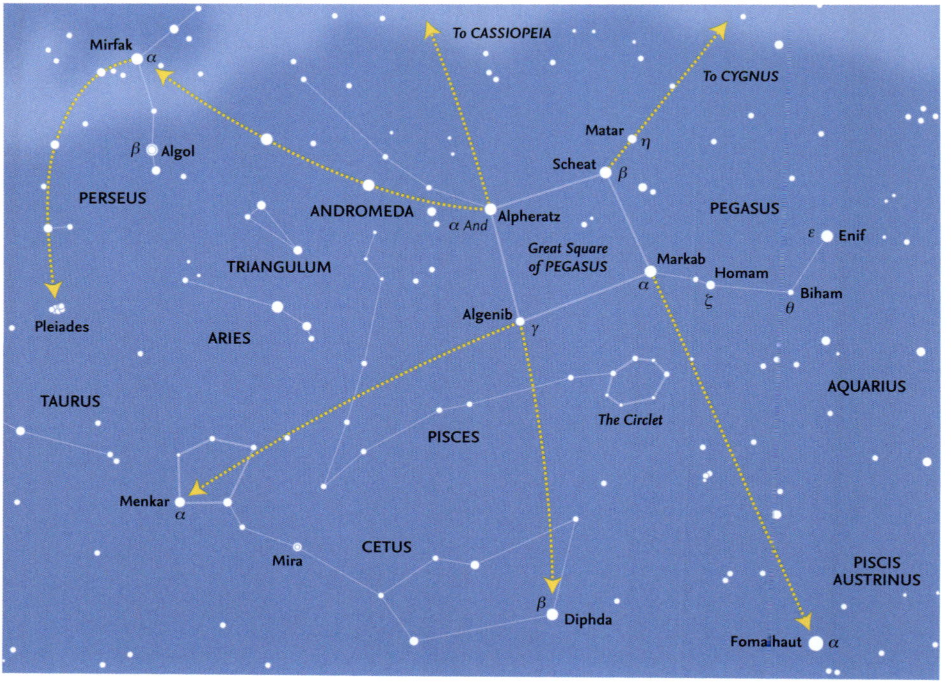

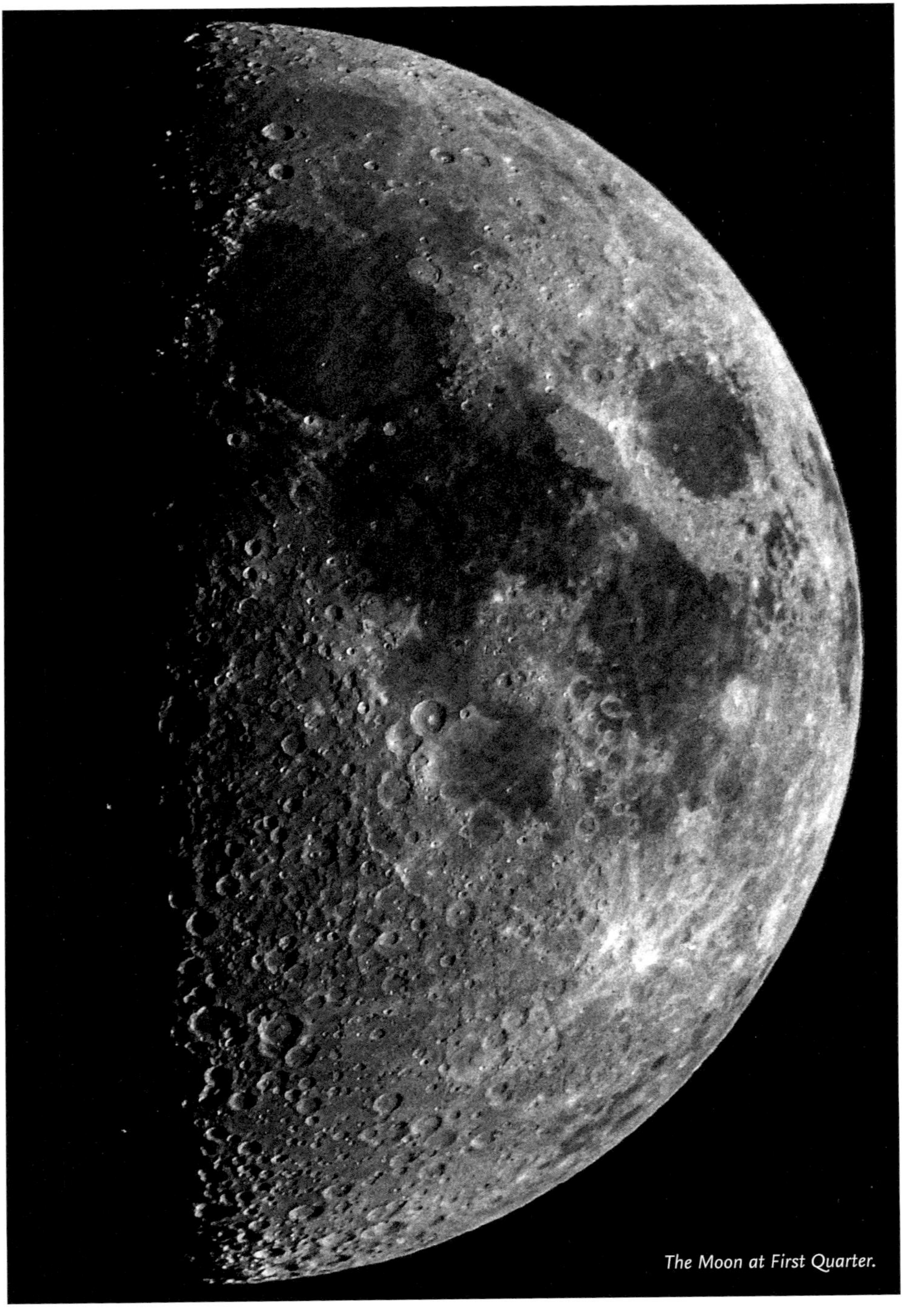

The Moon at First Quarter.

The Moon

The lunar phase cycle has a duration of 29.5 days and starts with the New Moon, when the near side of the Moon facing us is not illuminated by the Sun. The New Moon rises several hours after midnight and sets in the late afternoon or night; the First Quarter Moon rises in the late morning and sets around midnight; the Full Moon is above the horizon most of the night and the Last Quarter Moon rises around midnight and sets in the late morning.

Although the main features of the surface – the light highlands and the dark maria (seas) – may be seen with the naked eye, far more features may be detected with the use of binoculars or any telescope. The many craters are best seen when they are close to the terminator (the boundary between the illuminated and the non-illuminated areas of the surface), when the Sun rises or sets over any particular region of the Moon and the crater walls or central peaks cast strong shadows. These are best seen at crescent, quarter and gibbous phases. Many features become difficult to see at Full Moon, although this is the best time to see the bright ray systems surrounding certain craters. Accompanying the Moon map on the following pages is a list of prominent features, including the days in the lunar cycle when they are normally close to the terminator and thus easiest to see. The dates of visibility vary slightly through the effects of ***libration***. Because the Moon's orbit is inclined to the Earth's equator and also because it moves in an ellipse, the Moon appears to rock slightly from side to side (and nod up and down). Over time a total of 59 per cent of the lunar terrain can be observed. Areas near the limb (the edge of the Moon) may vary considerably in their location and visibility. This is easily noticeable with Mare Crisium (Sea of Crises, eastern limb of the Moon, in the western sky) and the craters Tycho (southern part of the Moon) and Plato (northern part). Another effect is that at crescent phases before and after New Moon, the normally non-illuminated portion of the Moon receives a certain amount of light, reflected from the Earth. This Earthshine may enable certain bright features (such as the craters Aristarchus, Kepler and Copernicus on the western limb of the Moon, in the eastern sky) to be detected.

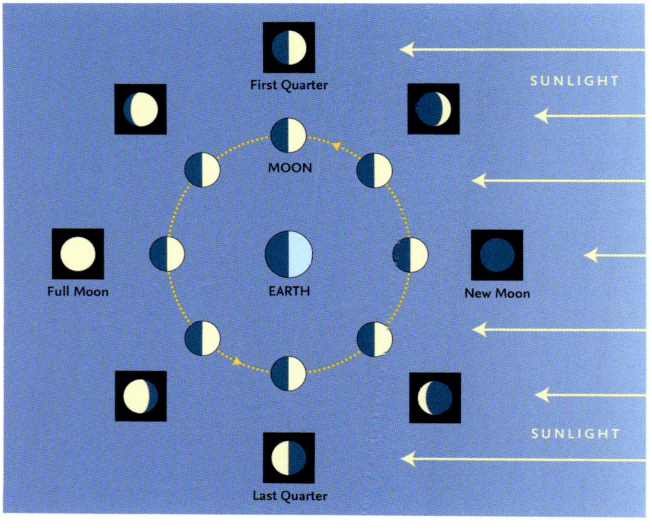

The Moon phases. *During its orbit around the Earth we see different portions of the illuminated side of the Moon's surface.*

Map of the Moon

Abulfeda	6:20
Agrippa	7:21
Albategnius	7:21
Aliacensis	7:21
Alphonsus	8:22
Anaxagoras	9:23
Anaximenes	11:25
Archimedes	8:22
Aristarchus	11:25
Aristillus	7:21
Aristoteles	6:20
Arzachel	8:22
Atlas	4:18
Autolycus	7:21
Barrow	7:21
Billy	12:26
Birt	8:22
Blancanus	9:23
Bullialdus	9:23
Bürg	5:19
Campanus	10:24
Cassini	7:21
Catharina	6:20
Clavius	9:23
Cleomedes	3:17
Copernicus	9:23
Cyrillus	6:20
Delambre	6:20
Deslandres	8:22
Endymion	3:17
Eratosthenes	8:22
Eudoxus	6:20
Fra Mauro	9:23
Fracastorius	5:19
Franklin	4:18
Gassendi	11:25
Geminus	3:17
Goclenius	4:18
Grimaldi	13-14:27-28
Gutenberg	5:19
Hercules	5:19
Herodotus	11:25
Hipparchus	7:21
Hommel	5:19
Humboldt	3:15
Janssen	4:18
Julius Caesar	6:20
Kepler	10:24
Landsberg	10:24
Langrenus	3:17
Letronne	11:25
Linné	6
Longomontanus	9:23

The numbers indicate the day or days in the lunar cycle when features are usually best visible.

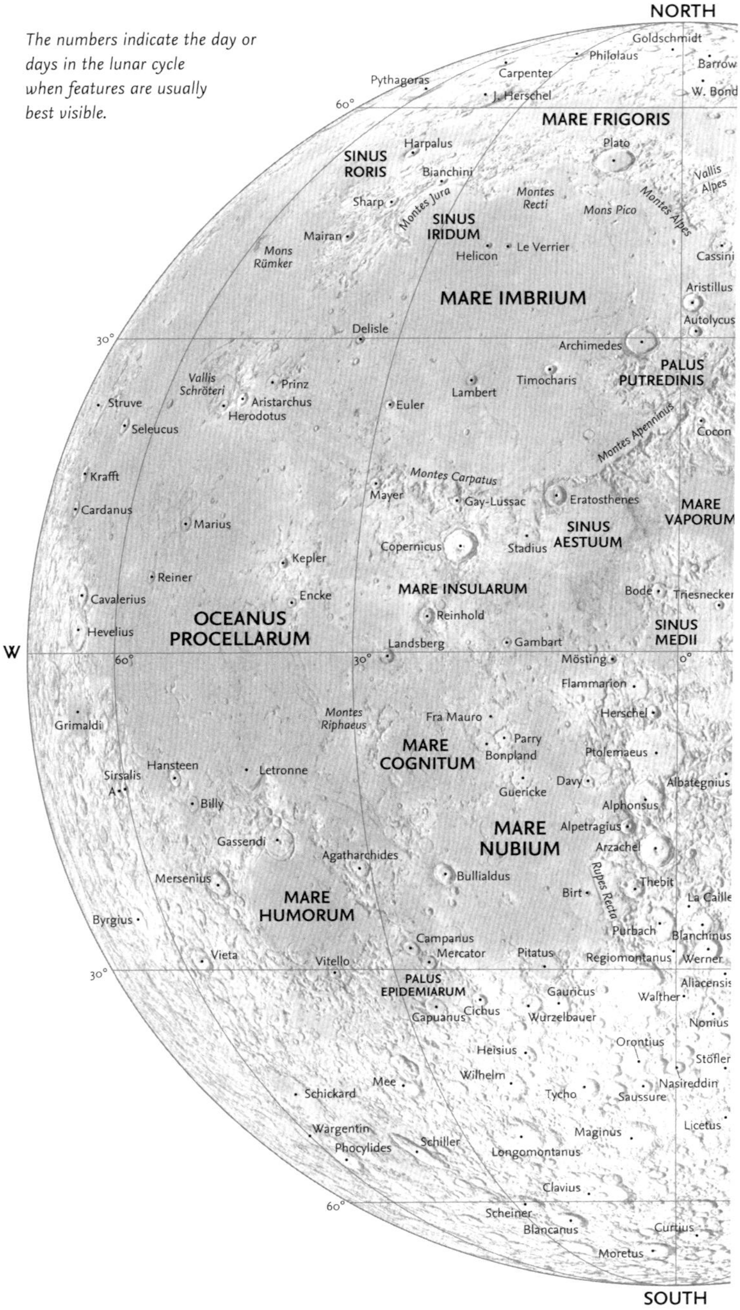

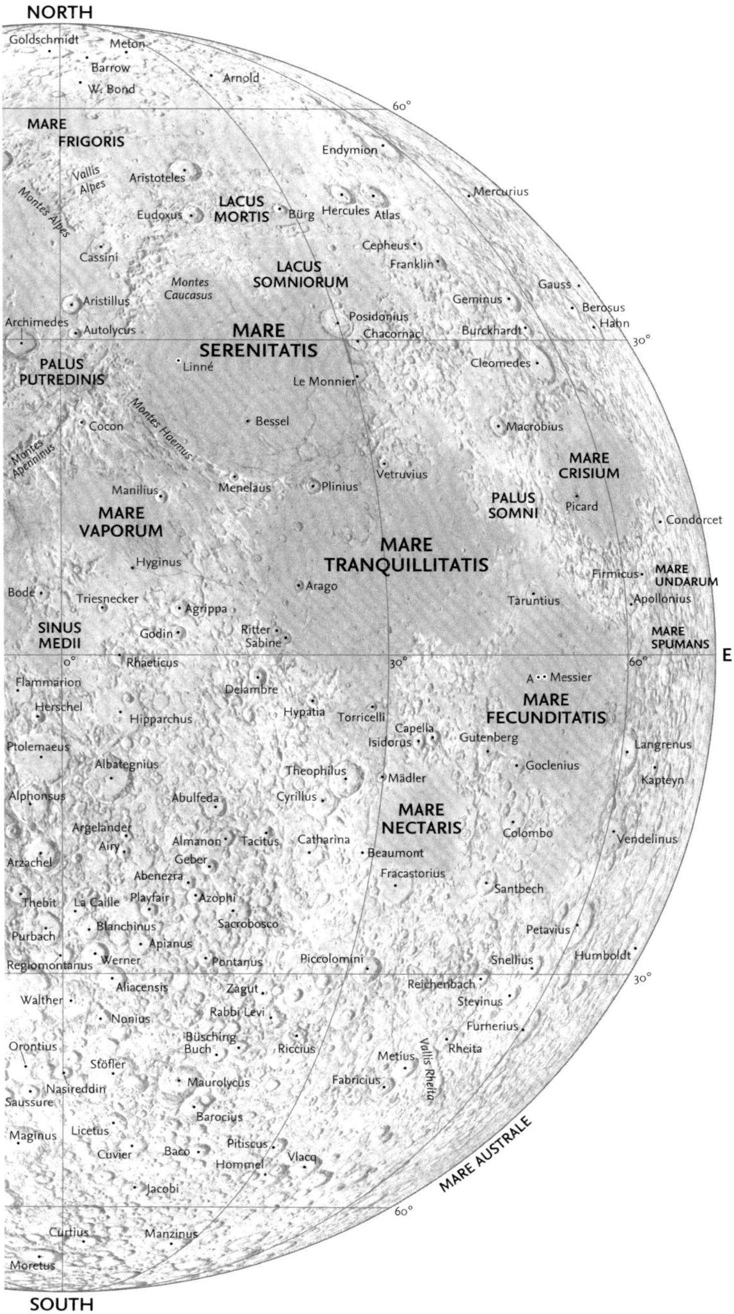
NORTH
Goldschmidt
Meton
Barrow
W. Bond
Arnold
60°
MARE FRIGORIS
Endymion
Vallis Alpes
Aristoteles
Montes Alpes
LACUS MORTIS
Eudoxus
Bürg
Hercules
Atlas
Mercurius
Cassini
Cepheus
Franklin
LACUS SOMNIORUM
Montes Caucasus
Aristillus
Gauss
Geminus
Berosus
Hahn
Posidonius
Archimedes
Autolycus
MARE SERENITATIS
Chacornac
Burckhardt
30°
PALUS PUTREDINIS
Linné
Le Monnier
Cleomedes
Montes Haemus
Cocon
Bessel
Macrobius
Montes Apenninus
MARE CRISIUM
Vetruvius
Manilius
Menelaus
Plinius
PALUS SOMNI
Picard
MARE VAPORUM
Condorcet
MARE TRANQUILLITATIS
Hyginus
Firmicus
MARE UNDARUM
Arago
Bode
Triesnecker
Agrippa
Taruntius
Apollonius
SINUS MEDII
Godin
Ritter
Sabine
MARE SPUMANS
E
0°
Rhaeticus
30°
60°
Flammarion
Messier
Delambre
Herschel
MARE FECUNDITATIS
Hipparchus
Hypatia
Torricelli
Capella
Isidorus
Gutenberg
Ptolemaeus
Langrenus
Albategnius
Goclenius
Theophilus
Mädler
Kapteyn
Alphonsus
Abulfeda
Cyrillus
MARE NECTARIS
Argelander
Colombo
Airy
Almanon
Tacitus
Catharina
Vendelinus
Beaumont
Geber
Arzachel
Abenezra
Fracastorius
Santbech
Thebit
La Caille
Playfair
Azophi
Blanchinus
Sacrobosco
Purbach
Petavius
Apianus
Regiomontanus
Werner
Pontanus
Piccolomini
Snellius
Humboldt
Aliacensis
Zagut
Reichenbach
30°
Walther
Stevinus
Nonius
Rabbi Levi
Furnerius
Büsching
Buch
Riccius
Vallis Rheita
Rheita
Orontius
Metius
Stöfler
Maurolycus
Fabricius
Nasireddin
Saussure
Barocius
Maginus
Licetus
MARE AUSTRALE
Pitiscus
Cuvier
Baco
Vlacq
Hommel
Jacobi
60°
Curtius
Manzinus
Moretus
SOUTH

Eclipses in 2027

Lunar eclipses

There are two lunar eclipses in 2027 and both are penumbral. The Moon will darken slightly during the eclipse as it passes through the lighter peripheral part of Earth's shadow; the dramatic blood red Moon effect is only seen with a total umbral lunar eclipse. The first penumbral eclipse takes place on 20 February, and it will be visible from Europe, Asia, western Australia, Africa, eastern regions of North America and South America. The maximum occurs at 23:12. The second penumbral eclipse takes place 17 August, maximum occurs at 7:13. The eclipse will be visible from the western tip of Europe, northeast parts of Asia, eastern regions of Australia, New Zealand, northwest Africa, North America and South America. The Moon will be below the horizon at this time in the UK and Ireland.

Solar eclipses

There are two solar eclipses in 2027. The first is an annular eclipse on 6 February with the maximum occurring at 15:59. The eclipse is visible from Chile and Argentina and it will appear as a partial eclipse from west and South Africa, Uruguay and Brazil. The duration of the annular eclipse is 7 minutes 51 seconds. There will be a total eclipse on 2 August visible from Spain, Morocco, Algeria, Libya, Egypt, Saudi Arabia, Yemen and Somalia. Maximum occurs at 10:06, the duration is 6 minutes 23 seconds. A partial solar eclipse will be visible from Europe, Africa, the Middle East, west and south Asia. From London, coverage of the Sun will reach 40% and the duration of the eclipse is 1 hour 56 minutes.

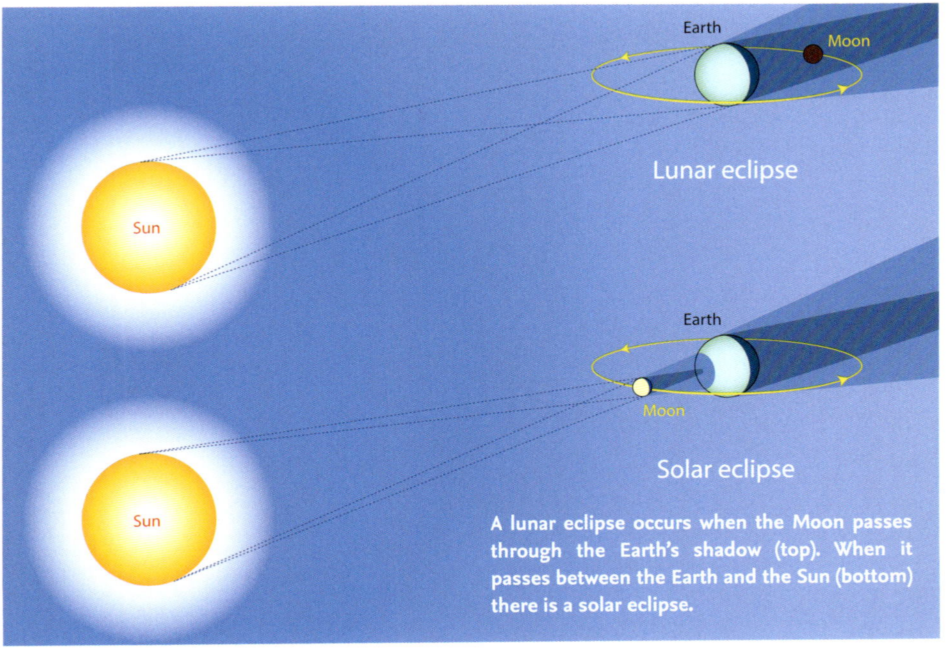

A lunar eclipse occurs when the Moon passes through the Earth's shadow (top). When it passes between the Earth and the Sun (bottom) there is a solar eclipse.

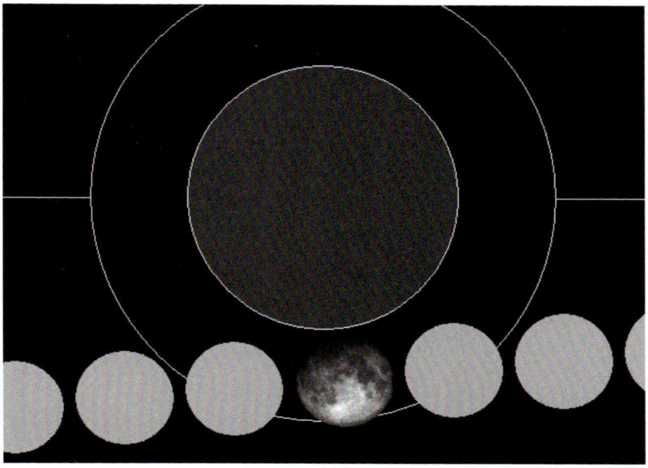

The path of the Moon as it passes through the perumbral shadow of the Earth on 20 February. Maximum eclipse occurs at 23:12 UT.

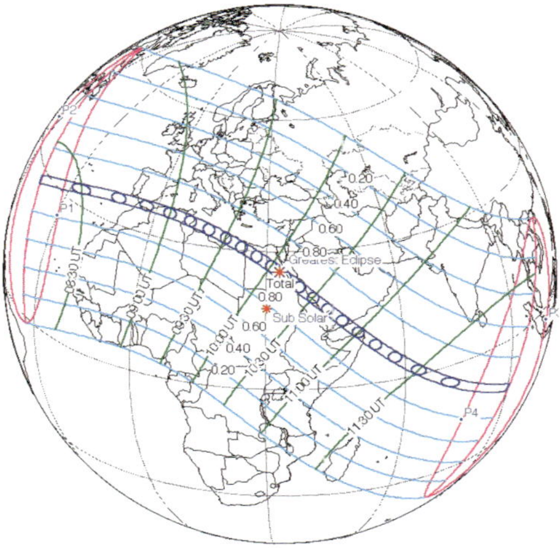

The path of the solar eclipse on 2 August.

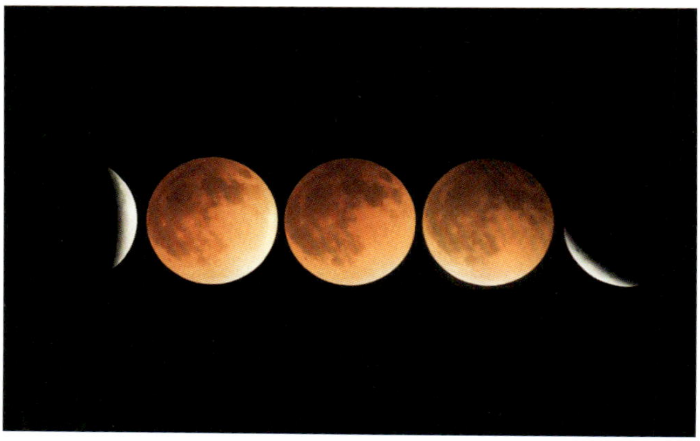

A sequence of five images showing the central stages of an eclipse, taken from Cascade-Siskiyou National Monument, Oregon in 2014.

The Planets in 2027

Mercury and Venus

Mercury reaches greatest eastern elongation three times during 2027 – it will be seen in the early evening of 3 February at mag. -0.5, 28 May (mag. 0.4) and 24 September (mag. 0.1, setting with Venus and Mars). Mercury switches to an appearance at dawn when it is at greatest western elongation on 17 March at mag. 0.2 (with Venus further south), 16 July at mag. 0.4 (with Venus closer to the Sun) and 4 November at mag. -0.5 (Jupiter sitting higher to the south). Its magnitude oscillates throughout the year – it will be brightest in April (-2.4) and faintest in October (6.3).

Venus starts the year leading the Sun in the dawn sky. It reaches greatest western elongation on 3 January appearing as the 'morning star' at mag. -4.5. Venus dims gradually throughout the year to mag. -3.9, and is faintest in May. Venus will appear in the evening sky beyond 11 August.

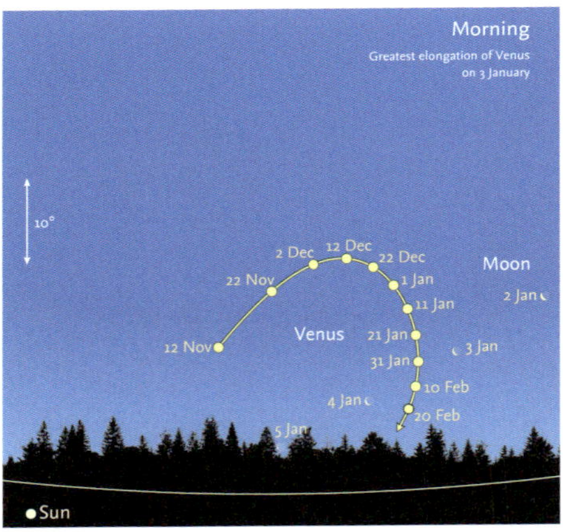

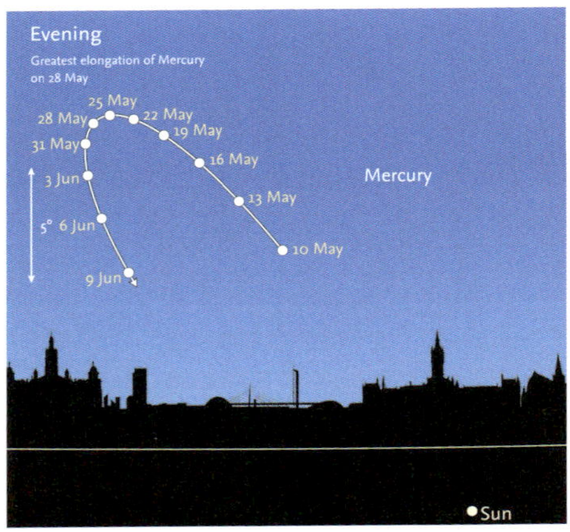

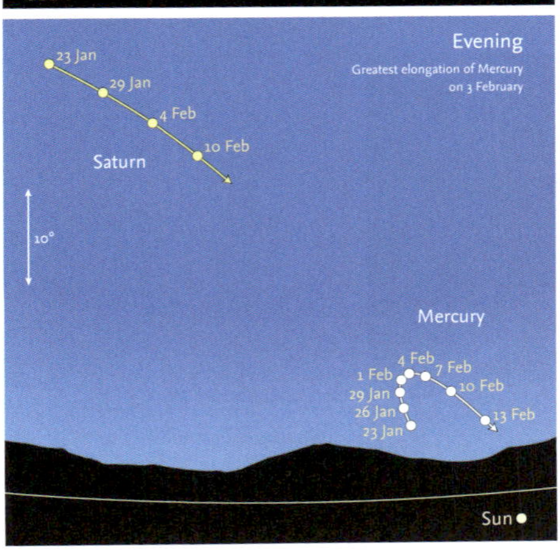

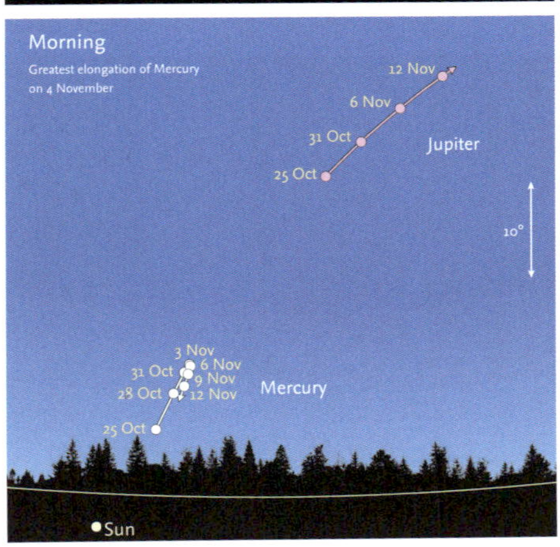

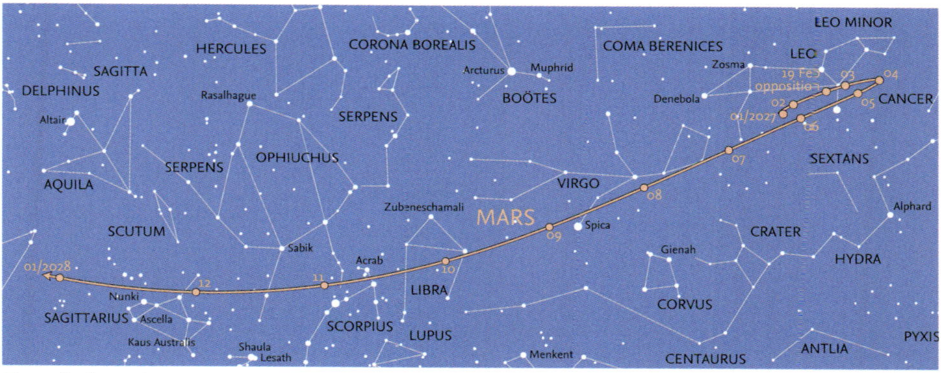

The path of Mars in 2027. Mars reaches opposition on 19 February. It enters retrograde motion on 10 January and reverts back to direct motion on 1 April. Mars is visible in the evening for most of the year.

Mars

Mars starts the year along the ecliptic in ***Leo***, rising late in the evening. It enters retrograde (westward) motion on 10 January, as the Earth overtakes it in its orbit around the Sun, reverting to direct (eastward) motion on 1 April. Mars moves into ***Virgo*** in July and then progresses into ***Libra*** in September. It shifts into ***Scorpio*** and then ***Ophiuchus*** in October, after which it settles into ***Sagittarius*** in November and eventually sets with the Sun in December.

Mars brightens to its peak magnitude of -1.3 in February while it sits in Leo. Its apparent size in the sky will increase as it reaches opposition on 19 February 2027. It gradually dims to its faintest in October (mag. 1.4) and it will brighten again towards the end of the year. As it takes longer to complete its orbit than the Earth, Mars does not come to opposition every year. Oppositions occur during a period of retrograde motion, when the planet appears to move westwards against the pattern of distant stars. Because of its eccentric orbit, which carries it at differing distances from the Sun (and Earth), not all oppositions of Mars are equally favourable for observation. The relative positions of Mars and the Earth are shown below. It will be seen that the opposition of 2018 was very close and thus favourable for observation. By comparison, opposition in 2027 will be at a far greater distance, so the planet will not appear as large as it did in 2018.

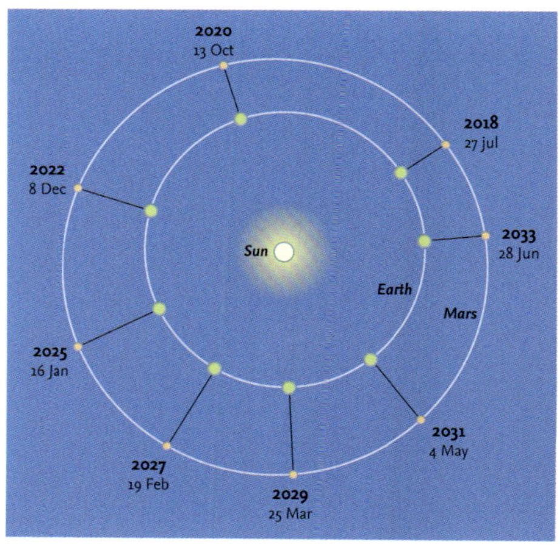

The oppositions of Mars between 2018 and 2033. The next opposition takes place on 19 February 2027.

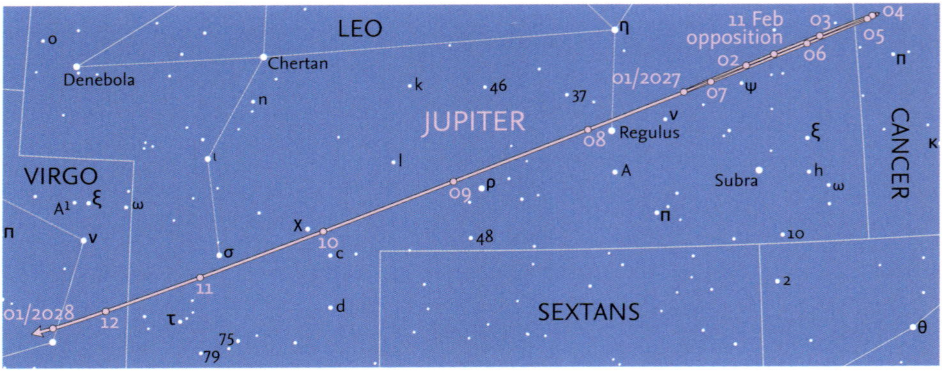

The path of Jupiter in 2027. Jupiter reaches opposition on 11 February. It ends retrograde motion on 13 April and re-enters retrograde westward motion in January 2028.

Jupiter and Saturn

Jupiter starts its journey in 2027 moving westward in retrograde motion along the ecliptic in ***Leo***. It reaches opposition on 11 February at peak brightness, mag. -2.6. Jupiter enters ***Cancer*** in March and ends retrograde motion on 13 April, staying in direct motion until January 2028. It moves back into Leo in May and sets with the Sun in late August, lost in the glare of the Sun and dimming to mag. -1.7. In October Jupiter reappears in the early morning sky and enters ***Virgo*** in November.

Jupiter's four largest satellites are readily visible in binoculars. Not all four are visible all the time, they are sometimes hidden behind the planet or invisible in front of it. Io, the closest to Jupiter, orbits in just under 1.8 days, and Callisto, the farthest away, takes about 16.7 days. In between are Europa (c. 3.6 days) and the largest, Ganymede (c. 7.2 days).

Saturn is in ***Cetus*** at the start of the year, it moves into ***Pisces*** in late February, visible in the early evening at its faintest (mag. 0.8). The ringed planet swings into the dawn sky in May and gradually pulls away from the Sun, brightening to its peak at opposition on 18 October (mag. 0.1). This will be a good time to look at the planet, its rings will be inclined by 13° to our line of sight (allowing observers to see more of the southern side). It enters retrograde motion on 9 August and creeps above the horizon before midnight from September. Saturn ends its retrograde motion in Pisces on 24 December.

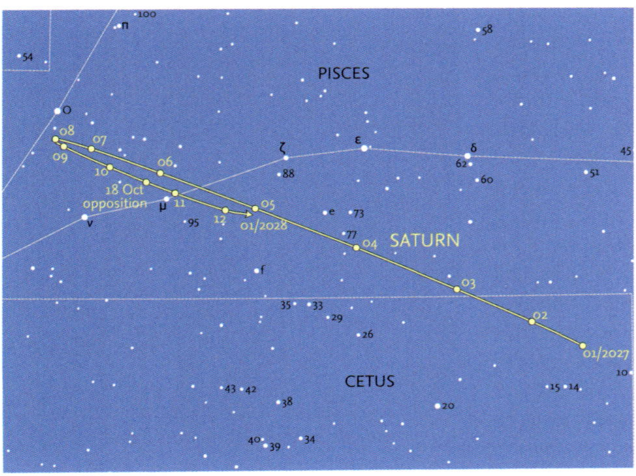

The path of Saturn in 2027. Saturn enters retrograde motion on 9 August, reverting back to direct eastward motion on 24 December. It reaches opposition on 18 October.

Uranus and Neptune

Uranus sits in ***Taurus*** for the whole of 2027. It leads the Sun in the dawn sky in June–July. It ends its retrograde motion on 8 February and re-enters retrograde on 15 September, moving westwards until 12 February 2028. Uranus will reach opposition on 30 November (mag. 5.6).

Neptune moved into ***Pisces*** in May 2022; it will stay in Pisces until August 2039. Neptune appears in the dawn sky from April, returning to the evening sky from September. It enters retrograde motion on 10 July at mag. 7.8. It will reach opposition on 28 September (mag. 7.7). Neptune reverts to direct motion on 15 December, fainter at mag. 7.8 and will continue moving eastwards until 11 July 2028.

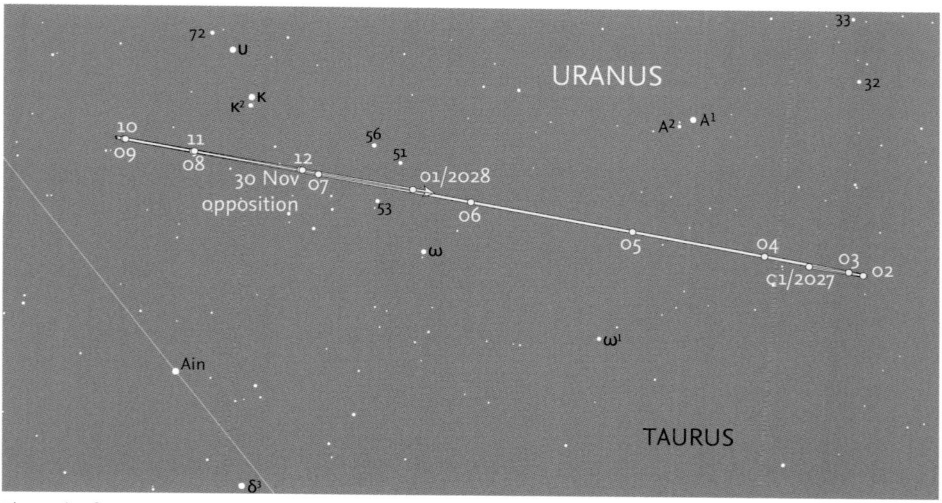

The path of Uranus in 2027 in Taurus. Uranus ends retrograde motion on 8 February, re-entering on 15 September. It reaches opposition on 30 November.

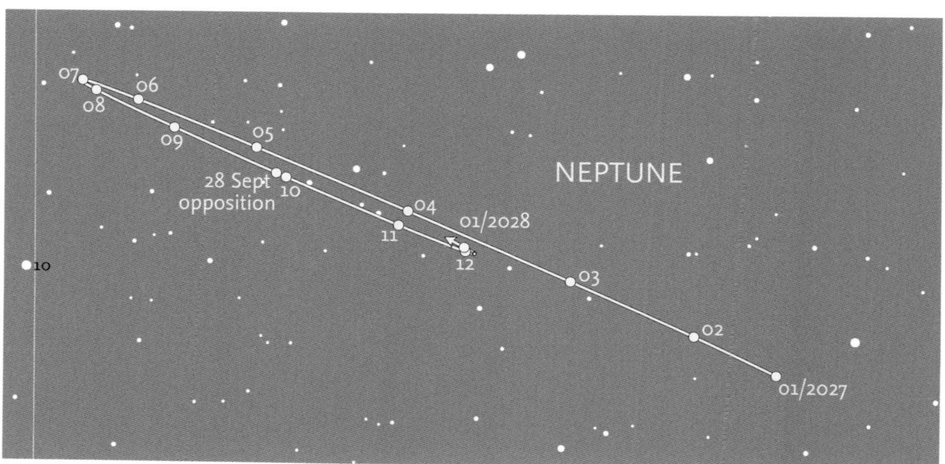

The path of Neptune in 2027 in Pisces. Neptune enters retrograde motion on 10 July, reaching opposition on 28 September. It reverts to direct eastward motion on 15 December.

Minor Planets in 2027

Three minor planets come to opposition and rise above magnitude 9.0 in 2027. They are closest to Earth at this point and reach their highest point in the sky on the meridian around midnight – this is the best time to see them. The year starts with the brightening of the dwarf planet ***(1) Ceres***, reaching a peak magnitude of 6.8 at opposition on 7 January 2027 when it will be at a distance of 1.63 AU (Astronomical Units, equivalent to the average distance from the Earth to the Sun). Named after the Roman goddess of agriculture, Ceres is the only dwarf planet that lies within the orbit of Neptune. It makes up 25% of the total mass of the asteroid belt (a vast ring of rocks between Mars and Jupiter). Ceres completes its orbit of the Sun in 4.6 years.

On 9 February the asteroid ***(27) Euterpe*** will lie in ***Leo***, reaching a peak magnitude of 8.9 at opposition, at a distance of 1.06 AU. On 21 September ***(29) Amphitrite*** will reach magnitude 8.8 in ***Pisces*** at opposition, passing within 1.45 AU of Earth.

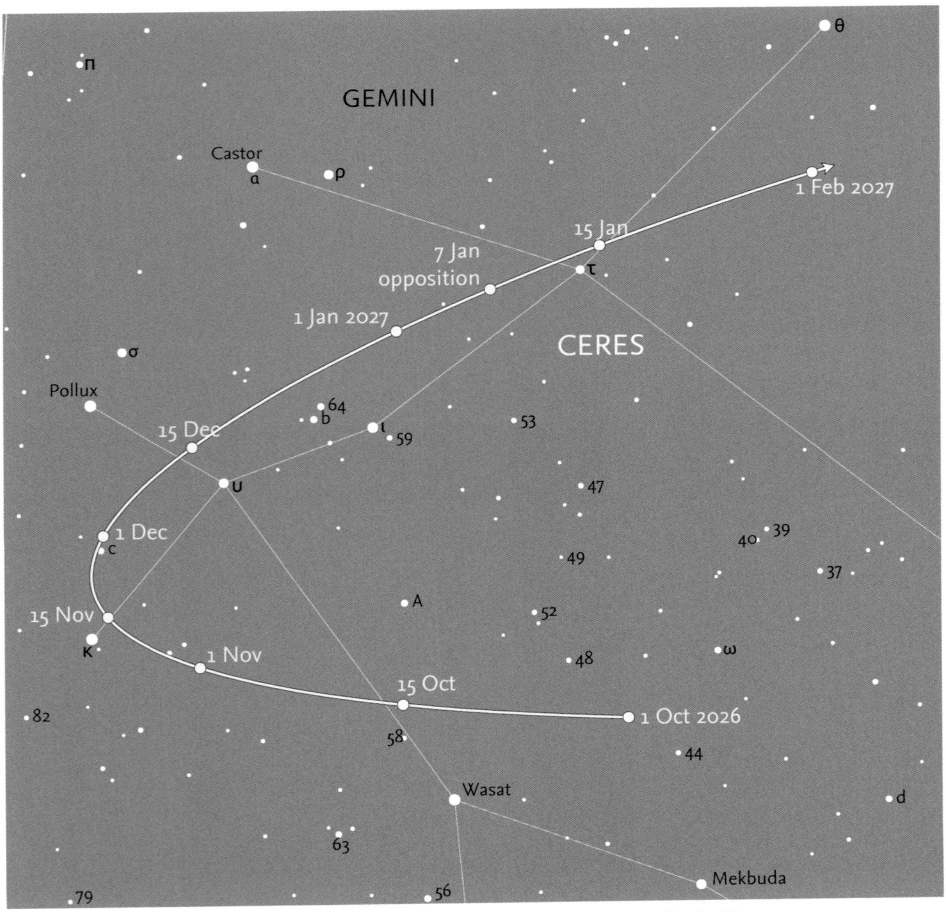

The path of (1) Ceres in 2027. (1) Ceres reaches opposition on 7 January, at magnitude 6.8.

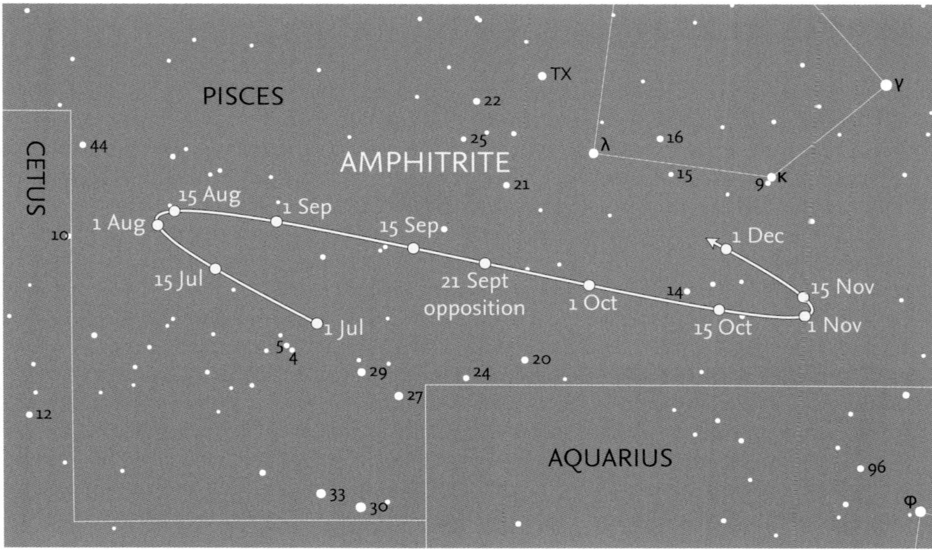

The path of (29) Amphitrite in 2027. (29) Amphitrite reaches opposition on 21 September, at magnitude 8.8.

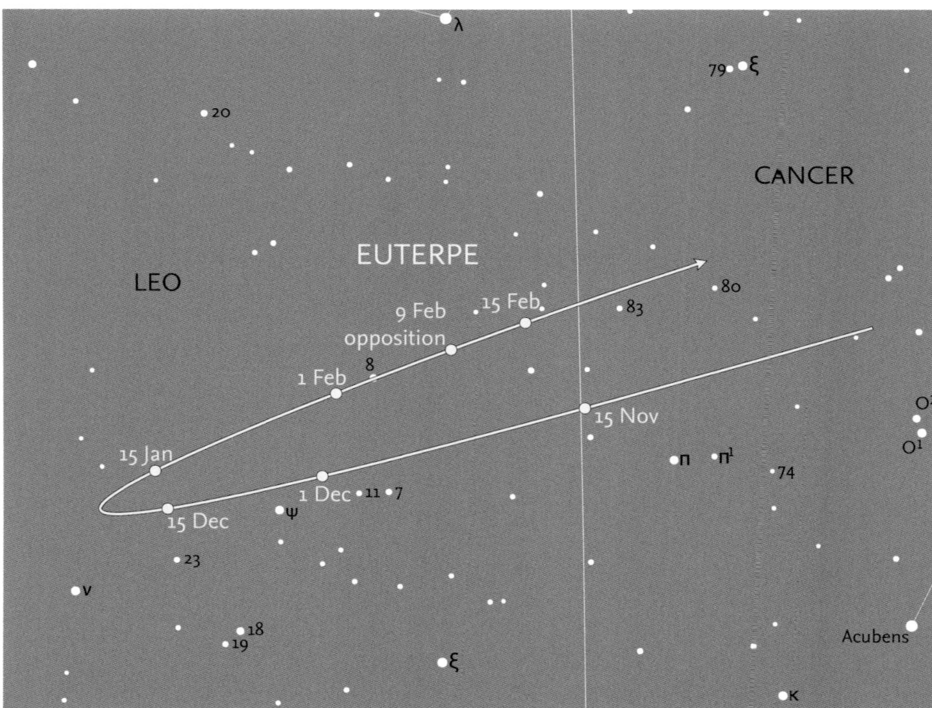

The path of (27) Euterpe in 2027. (27) Euterpe reaches opposition on 9 February, at magnitude 8.9.

Comets in 2027

Although comets may occasionally become very striking objects in the sky, their occurrence and particularly the existence or length of any tail and their overall magnitude are notoriously difficult to predict. Naturally, it is only possible to predict the return of periodic comets (whose names have the prefix 'P'). Many comets appear unexpectedly (these have names with the prefix 'C'). Bright, readily visible comets such as C/1995 Y1 Hyakutake, C/1995 O1 Hale-Bopp, C/2006 P1 McNaught or C/2020 F3 NEOWISE are rare (the names refer to the first observers or telescopes which discovered them). Most periodic comets are faint and only a very small number ever become bright enough to be easily visible with the naked eye or with binoculars.

Comet 2P/Encke will be visible early in the evening in January, moving through the constellations of Aquarius and Capricornus in the early part of the year and brightening to magnitude 5. It will make its closest approach to Earth (perigee) on 3 February 2027, passing within a distance of 0.62 AU. Comet 2P/Encke has a period of 3.3 years and the trail of debris left in its wake is the source of the Northern and Southern Taurid meteor showers, which occur in late October and November.

Comet 45P/Honda-Mrkos-Pajdusakova will progress through Orion, Taurus, Gemini, Cancer and Leo in August and September, brightening to around magnitude 8.5 as it approaches perihelion on 31 August. It will be very low in the sky just before sunrise. Also a short-period comet, it completes its orbit of the Sun in 5.3 years.

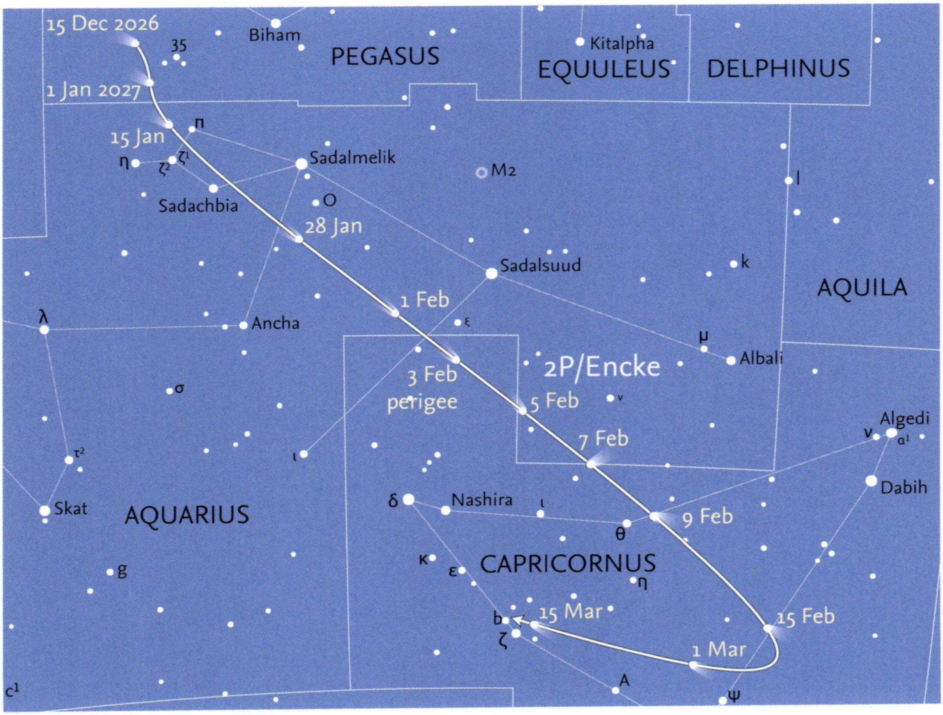

Comet 2P/Encke reaches magnitude 5 at its closest approach to Earth on 3 February.

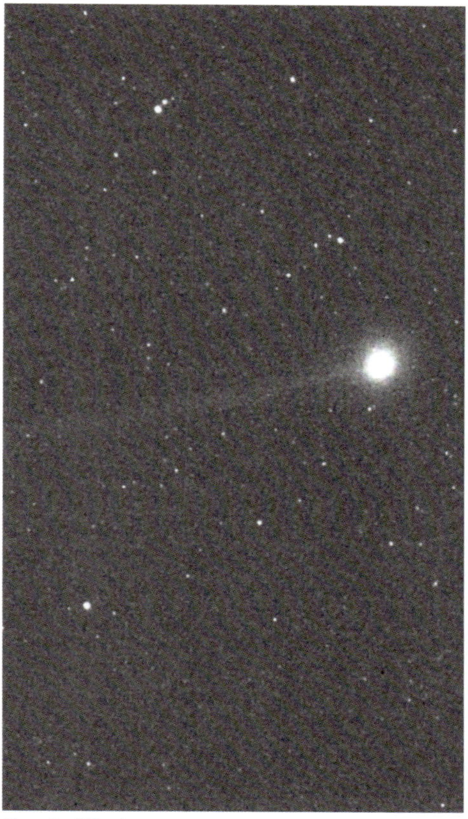

Comet 2P/Encke.

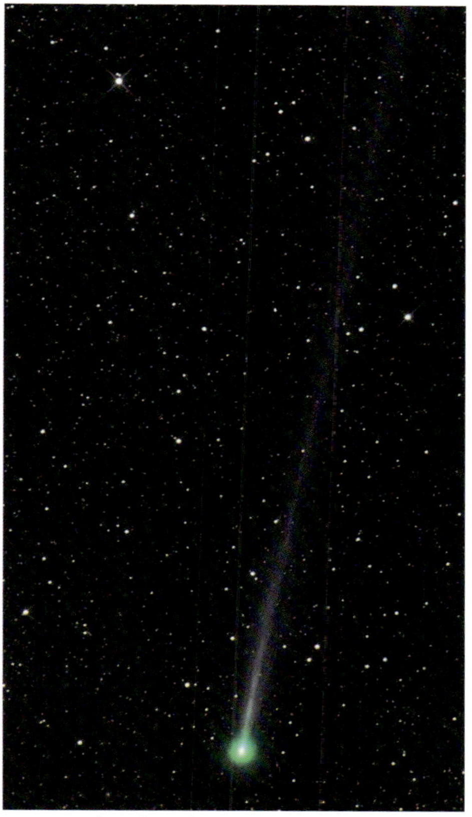

Comet 45P/Honda-Mrkos-Pajdusakova.

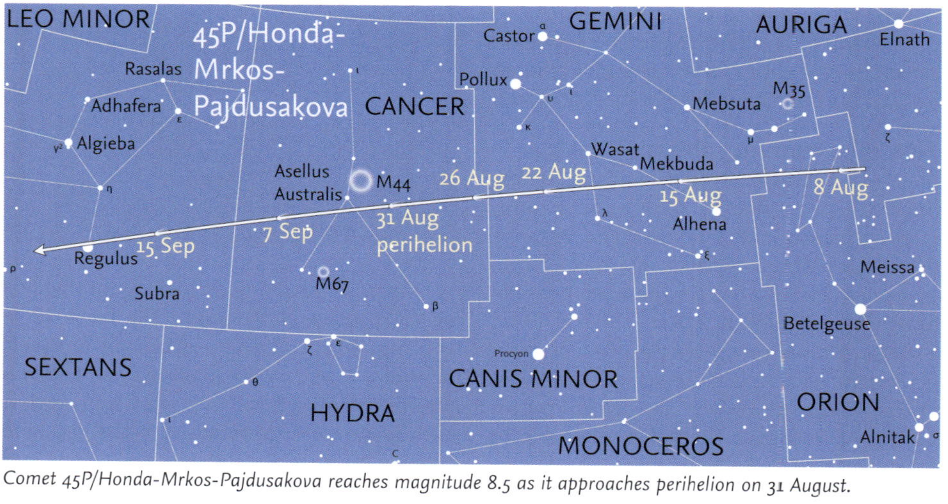

Comet 45P/Honda-Mrkos-Pajdusakova reaches magnitude 8.5 as it approaches perihelion on 31 August.

Introduction to the Month-by-Month Guide

The monthly charts

The pages devoted to each month contain a pair of charts showing the appearance of the night sky, looking north and looking south. The charts are drawn for the latitude of 50°N, so observers farther north will see slightly more of the sky on the northern horizon, and slightly less on the southern. Stars close to the horizon are always dimmed by atmospheric absorption (this is called extinction) and are also more affected by light pollution, so sometimes the faintest stars marked on the charts may not be visible.

The charts are drawn to show the appearance at 23:00 GMT for the 1st of each month. The same appearance will apply an hour earlier (22:00 GMT) on the 15th, and yet another hour earlier (21:00 GMT) at the end of the month (shown as the 1st of the following month). GMT is identical to the Universal Time (UT) used by astronomers around the world. In Europe, Summer Time is introduced in March, so the March charts apply to 23:00 GMT on 1 March, 22:00 GMT on 15 March, but 22:00 BST (British Summer Time) on 1 April. The change back from Summer Time (in Europe) occurs in October, so the charts for that month apply to 00:00 BST for 1 October, 23:00 BST for 15 October and 21:00 GMT for 1 November.

The charts may be used for earlier or later times during the night. To observe two hours earlier, use the charts for the preceding month; for two hours later, the charts for the next month.

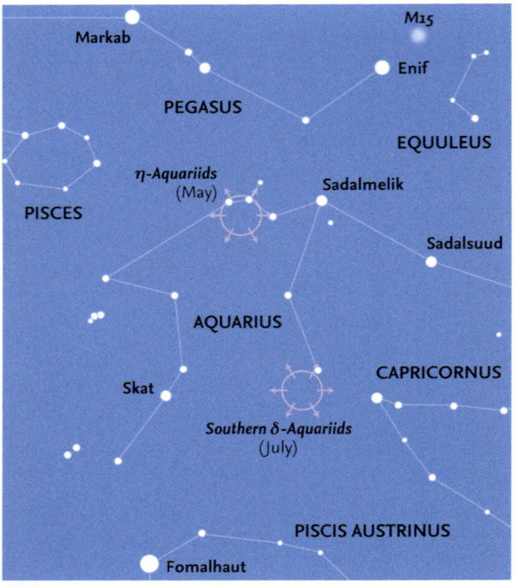

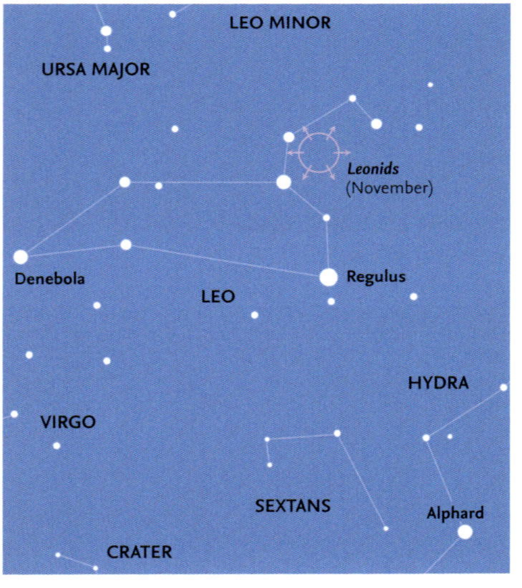

Tips for observing the night sky

A successful observing session requires planning, flexibility and patience. Check the weather forecast, sunset and moonrise and moonset times and choose a location where you have a clear view of the horizon wherever possible. Bring your star guide or app on a mobile device, a red torch to aid your night vision and a pair of binoculars and a tripod to keep them steady. You may be outside for a while – it's advised that you wear warm clothes, socks and shoes and bring hot drinks if it's a cold night.

If you are planning on looking for meteors or looking up at the night sky for a while you may wish to bring a deckchair or picnic blanket with you. When looking at faint objects (unlike the Moon) you will need to give your eyes 10–20 minutes to adapt to the dark and achieve night vision. If you are using your

Shower	*Dates of activity 2027*	*Date of maximum 2027*	*Possible hourly rate*
Quadrantids	December 12–January 12	January 3–4	120
April Lyrids	April 16–April 25	April 23	18
η-Aquariids	April 19–May 28	May 5–6	40
α-Capricornids	July 3–August 15	July 30	5
Southern δ-Aquariids	July 12–August 23	July 30	25
Perseids	July 17–August 24	August 12–13	150
α-Aurigids	August 28–September 5	September 1	6
Draconids	October 6–October 10	October 9	var.
Orionids	October 2–November 7	October 21–22	15
Southern Taurids	September 10–November 20	November 5	5
Northern Taurids	October 20–December 10	November 12–13	5
Leonids	November 6–November 30	November 17–18	15
Geminids	December 4–December 17	December 14	120
Ursids	December 17–December 26	December 22–23	10

phone, switch it to night vision mode and use a red torch to find your way around so that you maintain your night vision.

Astrophotography

You can capture the Moon and five planets visible to the naked eye or the brightest stars using your smartphone. The challenge is keeping your phone steady and gaining control over the focus and exposure (if the camera allows it). It is advised to attach the phone to a tripod, or you could use an adaptor to attach your phone to the eyepiece of a telescope – this will allow you to align it correctly with minimal effort. You can use the zoom function on your camera to change the magnification of the image.

There are many apps available that will provide more control of your camera settings and allow you to take better quality images: NightCap Pro, ProCam and AstroShader for iOS and Open Camera, DeepSkyCamera or Camera FV-5 for Android. For the Moon and bright planets, your camera should auto lock and focus, otherwise manually set the focus to infinity and keep the exposure short (a fraction of a second if it is a gibbous or Full Moon, one or two seconds for a thin crescent Moon and planets). You could control the shutter remotely via the volume control on wireless headphones, or set a timer on your phone (3 seconds delay). For fainter objects, increase the exposure and the ISO (the sensitivity of the camera detector); however, a higher ISO setting will introduce more noise. Trial and error will help you choose the best settings. Some apps allow images to be stacked; this means you can overlay images to increase brightness further. This is useful for the constellations, clusters such as the Pleiades and nebulous objects such as the Orion Nebula and Andromeda.

Meteors

Details of specific meteor showers are given in the months when they come to maximum. Radiants that lie below the horizon or in constellations not readily visible in that month are not marked. For this reason, special charts for the Eta (η) and Delta (δ) Aquariids (May and July, respectively) and the Leonids (November) are given here. Meteors from such showers may still be seen, because the most effective region for seeing meteors is some 40–45° from the radiant, and that area of sky may well be above the horizon. A table of the best meteor showers visible during the year is also given here. The rates given are those that an experienced observer might see under ideal conditions. Generally, the observed rates will be far lower.

The photographs

One or more photographs of constellations visible in certain specific months are included.

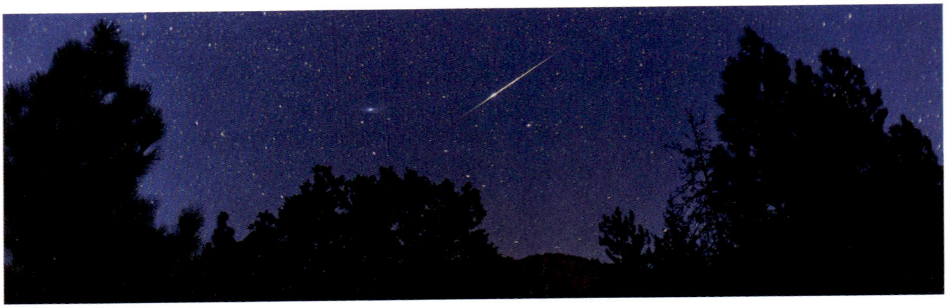

A Perseid fireball over Oregon.

It should be noted, however, that several factors affect observation: differences between the sensitivity of different individuals to faint starlight including colour, the degree to which they have become adapted to the dark, the local altitude of the star and the atmospheric conditions including local light pollution. Also, the exposure time can be altered on a digital camera, thus enhancing the intensity and colour of the object. Another factor to consider is the view of the object; some telescopes invert the image and in some cases the image is mirror-reversed (left and right are swapped). Each telescope will have a different optical configuration. Corrective optics can be used; however, the more optics are included, the more the light is attenuated as it passes through to the eyepiece. Binoculars provide an image with the correct orientation.

The Moon calendar

The Moon calendar shows the phase of the Moon for every day of the month. Also shown is the ***age*** of the Moon (the day in the ***lunation***), beginning at New Moon, which may be used to determine the best time for observation of specific lunar features.

The Moon

The section on the Moon includes details of any lunar or solar eclipses that may occur during the month. Similar information is given about any important occultations. Mainly, however, this section summarizes when the Moon passes close to planets or the five prominent stars close to the ecliptic. The dates when the Moon is closest to the Earth (at ***perigee***) and farthest from it (at ***apogee***) are shown in the monthly calendars, and only the nearest and farthest points during the year are mentioned.

The planets and minor planets

Brief details are given of the location, movement and brightness of the planets from Mercury to Saturn throughout the month. Mercury and Venus are normally easiest to see around either eastern or western elongation, in the evening or morning sky, respectively.

The dates at which the superior planets reverse their motion (from direct motion to ***retrograde***, and retrograde to direct) and of opposition (when a planet generally reaches its maximum brightness) are given. Jupiter and Saturn are normally easiest to see around opposition, which occurs every year. Mars, by contrast, moves relatively rapidly against the background stars and in some years never comes to opposition. The distant outer planets, especially Saturn, Uranus and Neptune, may spend most or all of the year in a single constellation.

Uranus is normally magnitude 5.7 to 5.9, and thus at the limit of naked-eye visibility under exceptionally dark skies, but bright enough to be readily visible in binoculars. Similar considerations apply to Neptune, it is always fainter (generally magnitude 7.8 to 8.0); however, it is still visible in most binoculars. The charts on page 25 show their positions during 2027.

In any year, few minor planets ever become bright enough to be detectable in binoculars. Just one, (4) Vesta, on rare occasions brightens sufficiently for it to be visible to the naked eye. Our limit for visibility (using binoculars or a telescope) is magnitude 9.0 and details and charts are given for those objects that exceed that magnitude during the year, normally around opposition. The background stars are shown to a fainter magnitude for better contrast. Minor planet charts for 2027 are on pages 26 and 27.

The ecliptic charts

Although the ecliptic charts are primarily designed to show the positions and motions of the major planets, they also show the motion of the Sun during the month. The light-tinted area shows the area of the sky that is invisible during daylight, but the darker area gives an indication of which constellations are likely to be visible at some time of the night.

The monthly calendar

For each month, a calendar shows details of significant events, including when planets are close to one another in the sky, close to the Moon, or close to any one of five bright stars that are spaced along the ecliptic. The times shown are given in Universal Time (UT), which is identical to Greenwich Mean Time (GMT).

The diagrams of interesting events

Each month, a number of diagrams show the appearance of the sky when certain events take place. However, the exact positions of celestial objects and their separations greatly depend on the observer's position on Earth. When the Moon is one of the objects involved, because it is relatively close to Earth, there may be very significant changes from one location to another. Close approaches between planets or between a planet and a star are less affected by changes of location, which may thus be ignored.

The diagrams showing the appearance of the sky are drawn for the latitude of London, so will be approximately correct for most of Britain and Europe. However, for an observer farther north (say, in Edinburgh), a planet or star listed as being north of the Moon will appear even farther north, whereas one south of the Moon will appear closer to it – or may even be hidden (occulted) by it. For an observer farther south than London, there will be corresponding changes in the opposite direction: for a star or planet south of the Moon, the separation will increase, and for one north of the Moon, the separation will decrease. This is particularly important when occultations occur, which may be visible from one location, but not another.

The details given regarding the positions of the various bodies should be used as a guide to their location. Similarly, dates and times are given, even if they fall in daylight, when the objects are likely to be completely invisible. However, such times do give an indication that the objects concerned will be in the same general area of the sky during both the preceding and the following nights.

Data used in this Guide

The data given in this Guide, such as timings and distances between objects, have been determined by the following sources: the Astronomical Applications Department of the US Naval Observatory; the Sky Events Calendar by Fred Espenak and Sumit Dutta (NASA's Goddard Space Flight Center); the NASA Jet Propulsion Laboratory (JPL) Horizons System; the International Astronomical Union's (IAU) Minor Planet Center; In-The-Sky.org.

Key to the symbols used on the monthy star maps.

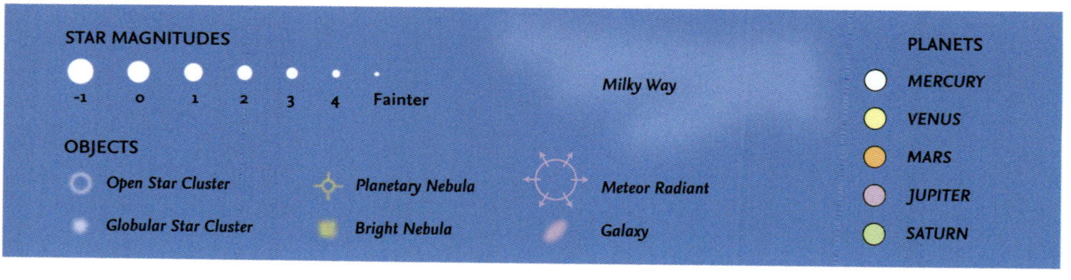

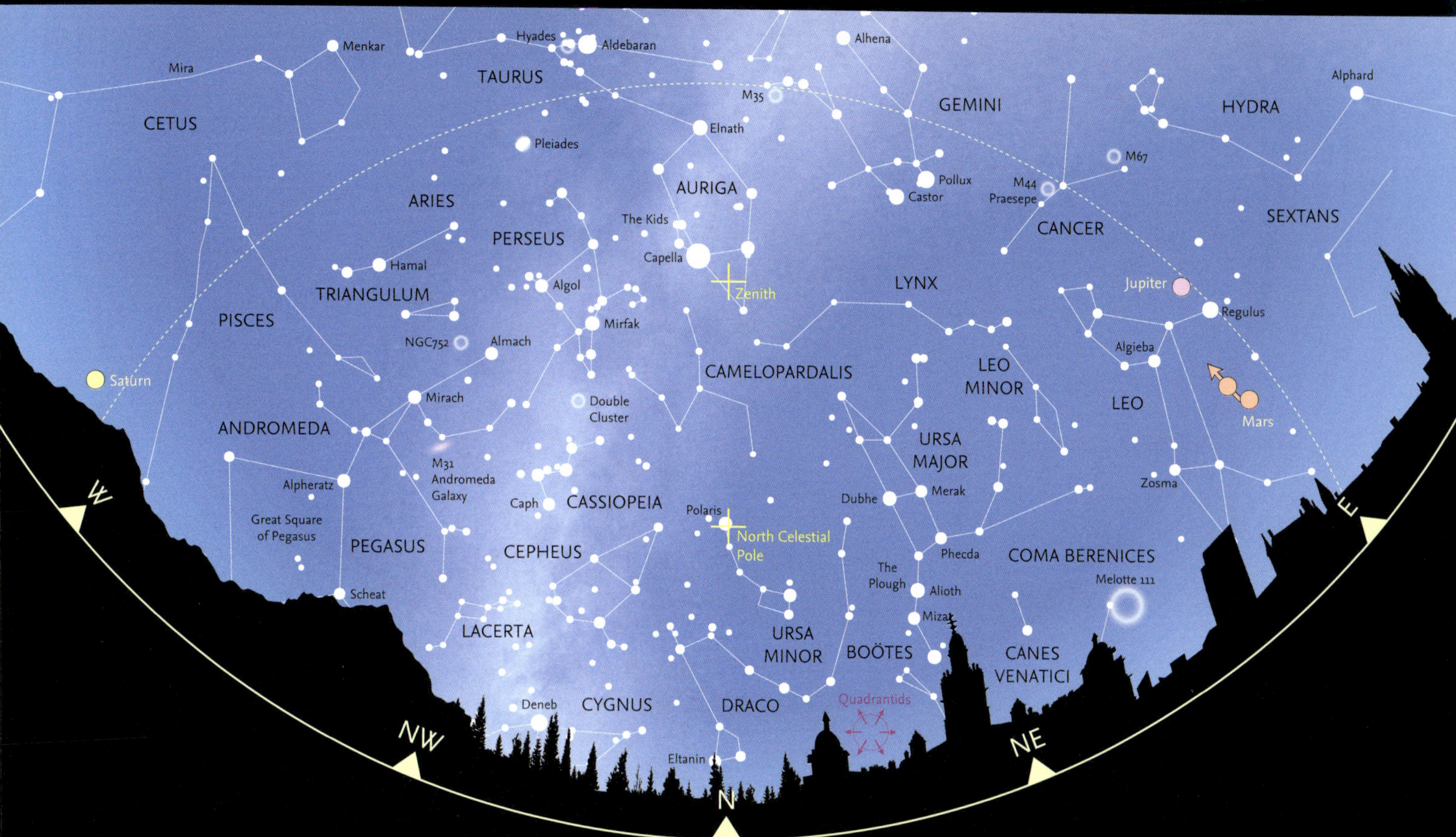

January – Looking North

1 January	23:00 GMT
15 January	22:00 GMT
1 February	21:00 GMT

January – Looking North

Most of the important circumpolar constellations are easy to see in the northern sky at this time of year. ***Ursa Major*** stands more or less vertically above the horizon in the northeast, with the zodiacal constellation of ***Leo*** rising in the east. To the north, the stars of ***Ursa Minor*** lie below ***Polaris*** (the Pole Star). The head of ***Draco*** is low on the northern horizon, the stars may be difficult to see unless observing conditions are good. Both ***Cepheus*** and ***Cassiopeia*** are readily visible in the northwest, and even the faint constellation of ***Camelopardalis*** is high enough in the sky for it to be easily visible.

Near the zenith is the constellation of ***Auriga*** (the Charioteer), with brilliant ***Capella*** (α Aurigae), directly overhead. Slightly to the west of Capella lies a small triangle of fainter stars, known as the Kids. (Ancient mythological representations of Auriga show him carrying two young goats.) Together with the northernmost bright star in ***Taurus, Elnath*** (β Tauri), the body of Auriga forms a large pentagon on the sky, with the Kids lying on the western side. Farther down towards the west are the constellations of ***Perseus*** and ***Andromeda***, and the Great Square of ***Pegasus*** is approaching the horizon.

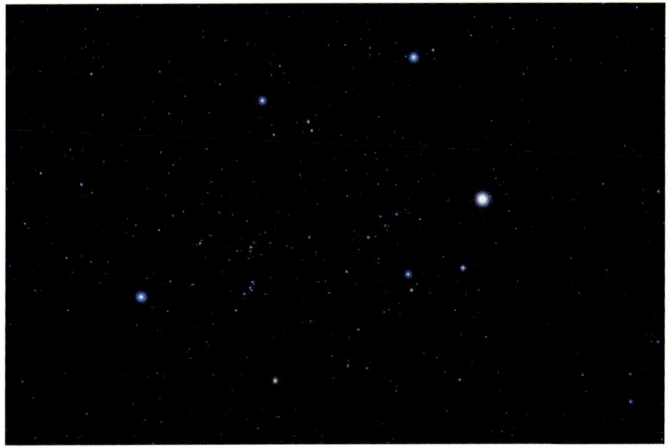

The constellation Auriga and its brightest star, Capella, the sixth brightest star in the night sky.

Meteors

The ***Quadrantid*** meteor shower begins on 12 December 2026 and continues until 12 January 2027. It is one of the strongest and most consistent meteor showers. It comes to maximum on the night of 3 and 4 January, best seen just after midnight while the Moon is far below the horizon. The meteors are bright, bluish- or yellowish-white, and may reach a maximum rate of 120 per hour. The parent object is minor planet 2003 EH_1.

The shower is named after the former constellation ***Quadrans Muralis*** (the Mural Quadrant), an early form of astronomical instrument. The Quadrantid radiant, marked on the chart, is now within the northernmost part of ***Boötes***, roughly halfway between θ Boötis and τ Herculis.

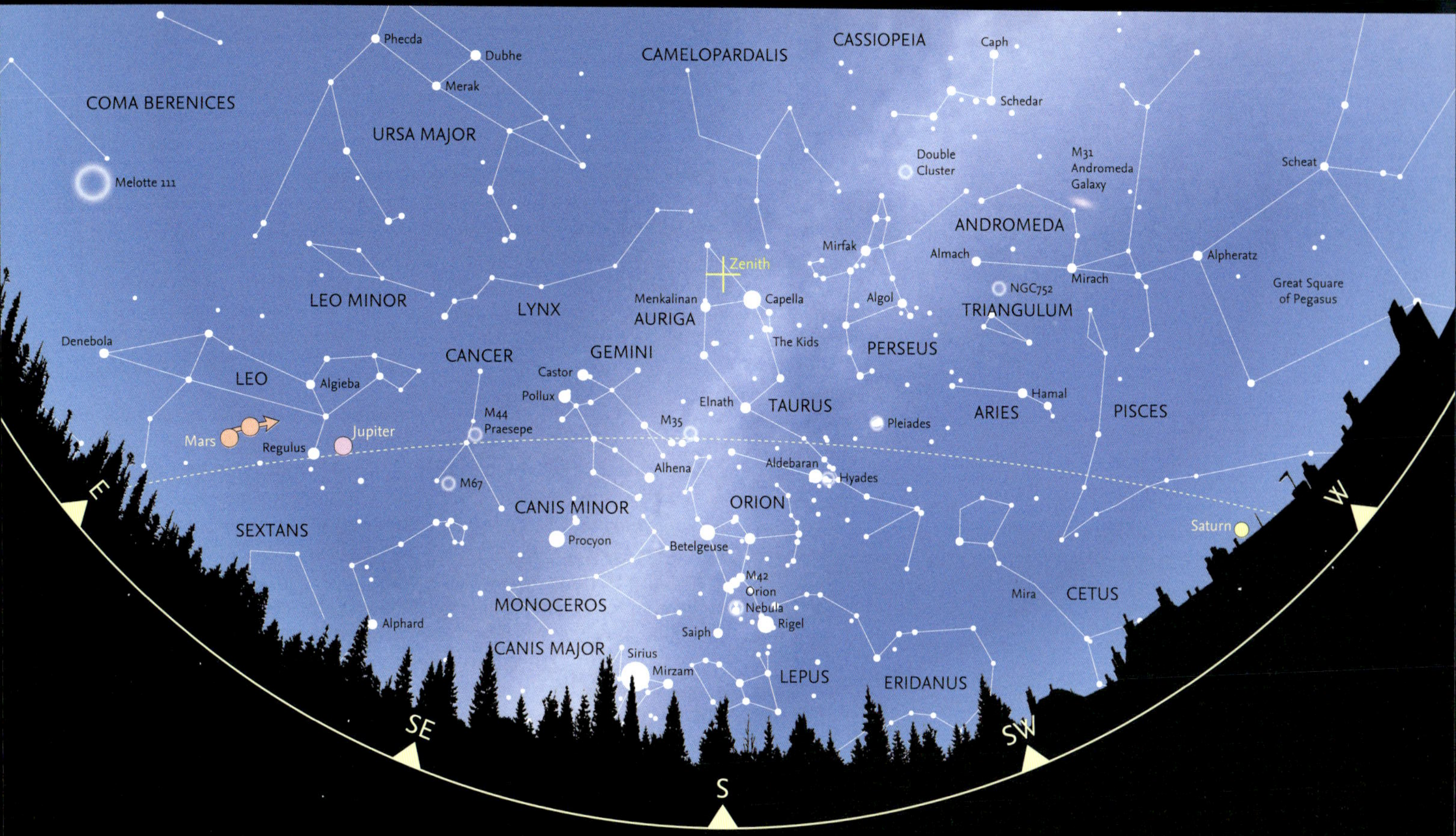

January – Looking South

1 January	23:00 GMT
15 January	22:00 GMT
1 February	21:00 GMT

January – Looking South

The southern sky is dominated by ***Orion***, prominent during the winter months, visible at some time during the night. It is highly distinctive, with a line of three stars that form Orion's Belt. To most observers, the bright star ***Betelgeuse*** (α Orionis), shows a reddish tinge; it is a red supergiant star, in contrast to the brilliant bluish-white supergiant ***Rigel*** (β Orionis). The three stars of the belt lie directly south of the celestial equator. A vertical line of three 'stars' forms the 'sword' that hangs south of the 'belt'. With good viewing, the central 'star' appears as a hazy spot, even to the naked eye, and is actually the ***Orion Nebula***, a star formation region. Binoculars reveal the four stars of the ***Trapezium*** cluster, which illuminate the nebula, as well as more of the nebula.

The Belt points up to the northwest towards ***Taurus*** (the Bull) and orange-tinted ***Aldebaran*** (α Tauri). Close to Aldebaran, there is a conspicuous 'V' of stars, called the ***Hyades*** cluster. Aldebaran lies much closer to the Earth than the Hyades, it just so happens to share the same line of sight. Farther along, the same line from Orion passes below a bright cluster of stars, the ***Pleiades***, or Seven Sisters. Even the smallest pair of binoculars reveals this as a beautiful group of bluish-white stars. Another conspicuous star in Taurus, ***Elnath*** (β Tauri), lies directly above Orion and forms a big 'V' with Aldebaran.

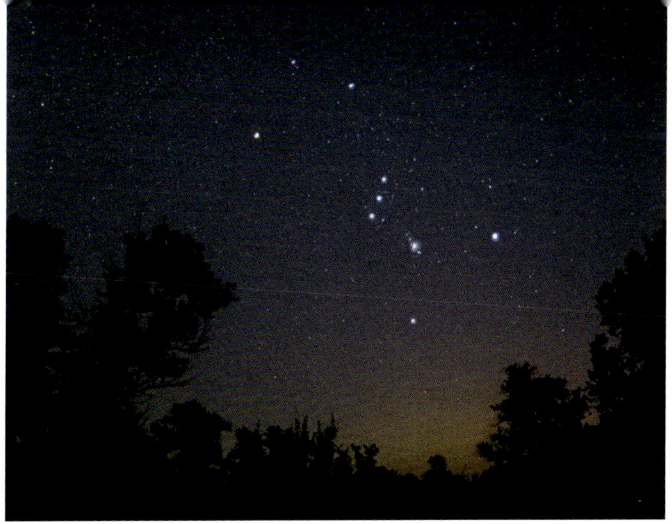

Orion the Greek hunter.

The Moon's phases for January 2027

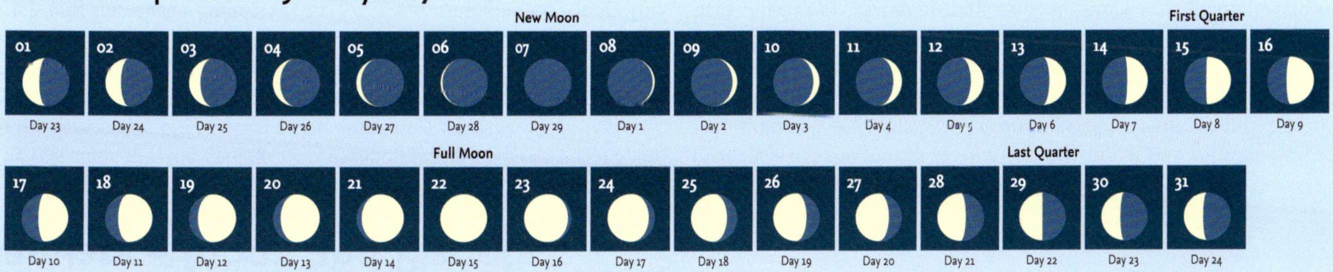

January – Moon and Planets

The Moon

The year starts with the waning crescent Moon sitting 2.8°S of the star ***Spica*** in the early hours of New Year's Day. Three days later the Moon lies only 0.3°S of the red-orange star ***Antares***, visible just before sunrise. On 7 January the New Moon reaches its farthest point from Earth in the year at a distance of 406,610 km. First Quarter Moon occurs on 15 January and on 18 January the young star cluster ***Pleiades*** lies 1.1°S of the waxing gibbous Moon. The Moon is at its closest point to Earth in the year (357,285 km) on 21 January, that night it lies 4.4°S of the star ***Pollux*** and becomes Full the following day. On 24 January, a few hours before sunrise, the waning gibbous Moon sits 1.5°S of ***Jupiter*** (mag. -2.5) and 1.5°S of ***Regulus***. On 28 January the Moon passes close to ***Virgo***'s brightest star ***Spica*** again, appearing 2.9°S after midnight. The day after brings a Last Quarter Moon and the month ends with Antares sitting 0.4°N of the waning crescent Moon.

The Planets

Mercury starts the month at solar conjunction, it sets with the Sun and then gradually trails the Sun, becoming visible at sunset and dimming from mag. -1.3 to -0.8 at the end of January. ***Venus*** leads the Sun at dawn at the start of the year, reaching eastern elongation on 3 January at mag. -4.5 and dimming to mag. -4.3 at the end of the month. ***Mars*** rises after 21:00 in ***Leo*** at mag. -0.1, it enters retrograde (westward) motion on 10 January and completes the month at mag. -0.9. ***Jupiter*** lies in ***Leo***, rising ahead of Mars and best seen later in the evening. Our largest planet brightens from mag. -2.4 to mag. -2.5, passing 1.5°N of the waning gibbous Moon on 24 January. ***Saturn*** lies in ***Cetus*** during the month. It's visible until an hour before midnight and dims from mag. 0.7 to mag. 0.8. ***Uranus*** sits in ***Taurus*** most of the night, starting at mag. 5.6 and dimming to 5.7. ***Neptune*** is in ***Pisces*** at mag. 7.8, but is not visible after 23:00.

The path of the Sun and the planets along the ecliptic in January.

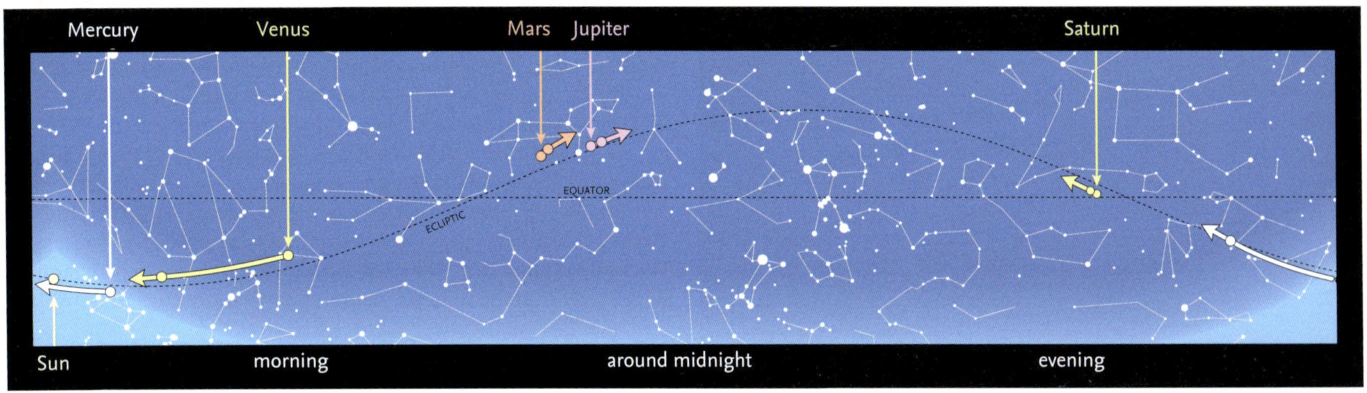

Calendar for January

01	00:29	Spica 2.8°N of Moon
03	02:32	Earth at perihelion (147,099,586 km = 0.9833 AU)
03	18:21	Venus at greatest elongation (47.0°W, mag. -4.5)
03–04		Quadrantid meteor shower maximum
04	16:34	Antares 0.3°N of the Moon
07	08:10	Moon at apogee = 406,610 km
07	20:24	New Moon
10	12:58	Mars enters retrograde motion
15	20:34	First Quarter Moon
18	08:46	Pleiades 1.1°S of the Moon
21	21:49	Moon at perigee = 357,285 km
21	23:18	Pollux 4.4°N of the Moon
22	12:17	Full Moon
24	00:24	Jupiter (mag. -2.5) 1.5°N of the Moon
24	09:19	Regulus 1.5°N of the Moon
28	07:55	Spica 2.9°N of the Moon
29	10:55	Last Quarter Moon
31	22:48	Antares 0.4°N of the Moon

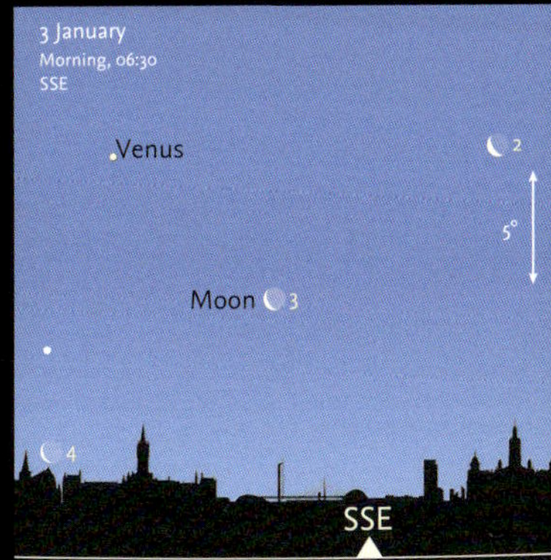

3 January • *The waning crescent Moon passes by Venus.*

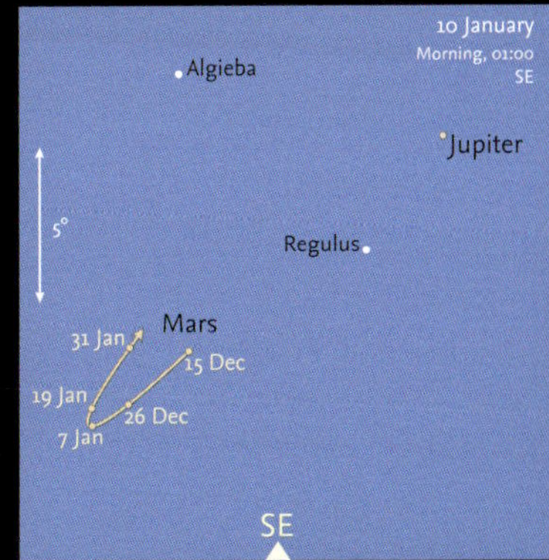

10 January • *Mars enters retrograde (westward) motion, Jupiter and Regulus are close by.*

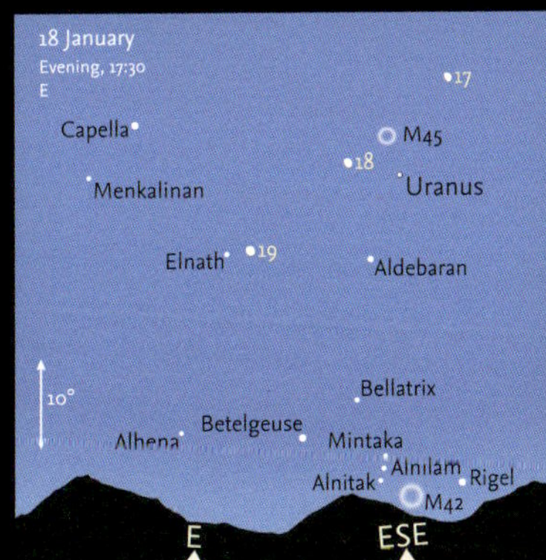

18 January • *The waxing gibbous Moon close to the Pleiades (M45) and Uranus.*

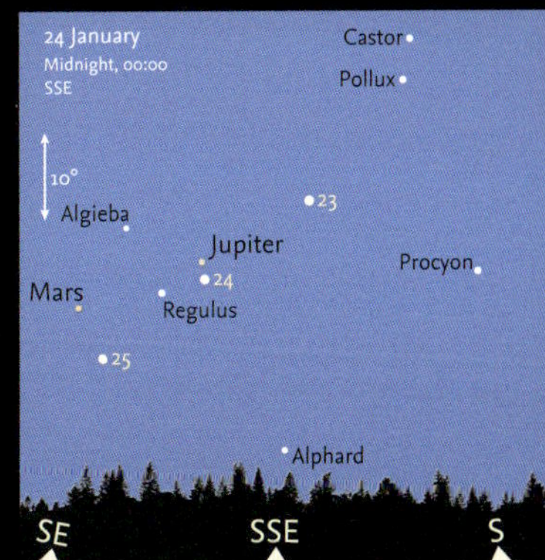

24 January • *Jupiter and the waning gibbous Moon together in the sky, Mars further east.*

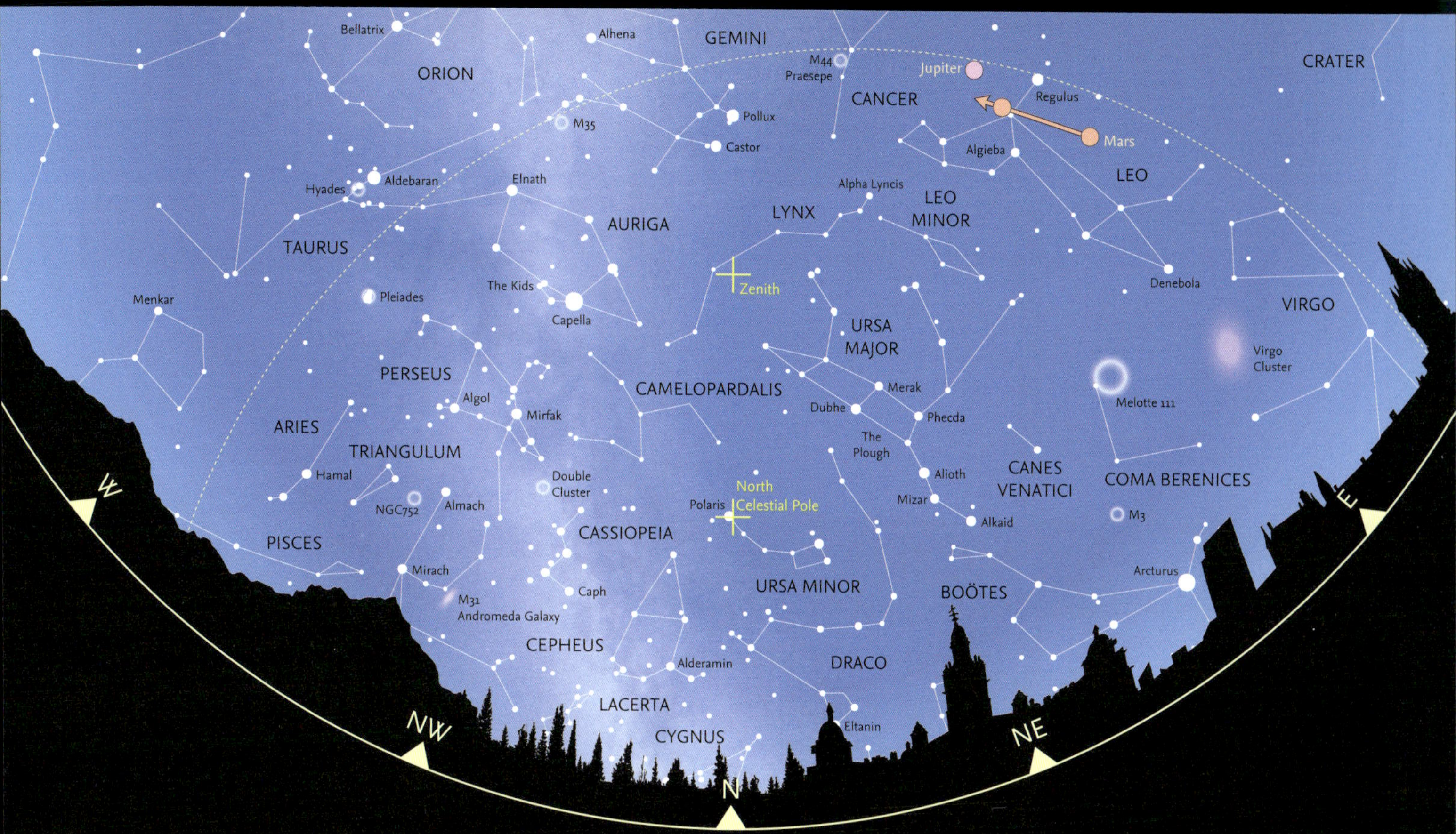

February – Looking North

1 February	23:00 GMT
15 February	22:00 GMT
1 March	21:00 GMT

February – Looking North

The denser band of stars of the Milky Way, first observed through a telescope by Galileo Galilei in 1610, run through the northern and western sky. The band stretches from ***Cygnus***, low on the northern horizon, through ***Cassiopeia***, ***Perseus*** and ***Auriga*** and reaching the southern sky via ***Gemini*** and ***Monoceros*** (see chart on page 42). Although not as readily visible as the denser star clouds of the summer Milky Way, on a clear night so many stars may be seen that even a distinctive constellation such as Cassiopeia is not immediately obvious.

The head of ***Draco*** is now higher in the sky and easier to recognize. The blue supergiant ***Deneb*** (α Cygni), the brightest star in Cygnus, may just be visible, almost due north at midnight. Later in the night the constellation of ***Boötes*** the herdsman, resembling a kite or diamond, clears the horizon. The orange-tinted red giant ***Arcturus*** (α Boötis) marks an apex. At magnitude -0.05, it is the brightest star in the northern hemisphere of the sky. ***Coma Berenices*** (Berenice's Hair) is now well above the horizon in the east. Within the constellation is a faint open star cluster of around 40 stars called ***Melotte 111*** or the Coma Star Cluster (not to be confused with the important Coma Cluster of galaxies, Abell 1656, mentioned on page 55).

On the other side of the sky, in the northwest, most of the constellation of ***Andromeda*** is still easily seen, although ***Alpheratz*** (α Andromedae), the star that forms the northeastern corner of the Great Square of ***Pegasus*** – even though it is actually part of Andromeda – is approaching the horizon and becoming more difficult to detect. High overhead, at the zenith, try to make out the very faint constellation of ***Lynx***. It was introduced in 1687 by the famous astronomer Johannes Hevelius to fill the largely blank area between Auriga, Gemini and ***Ursa Major***, and is reputed to be so named because one needed the eyes of a lynx to detect it. The brightest star, α Lyncis has a magnitude of 3.1.

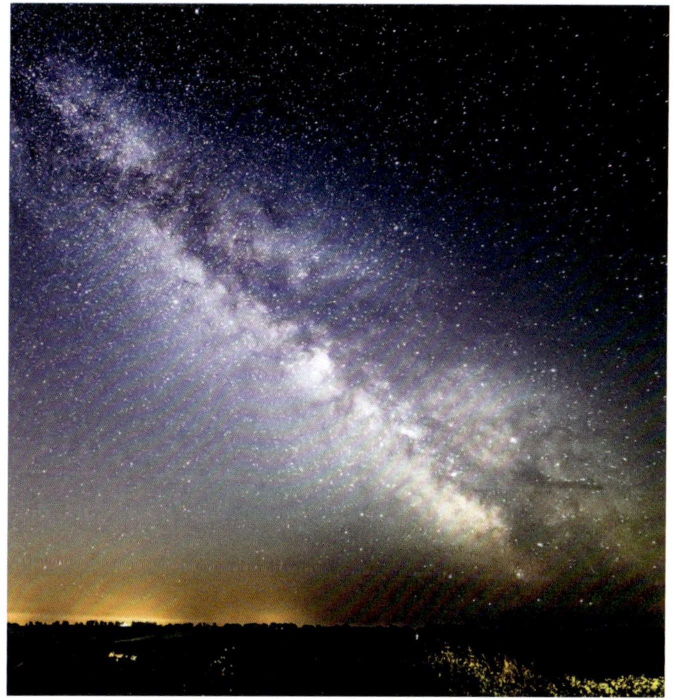

The Milky Way: the darker regions are dust clouds blocking starlight.

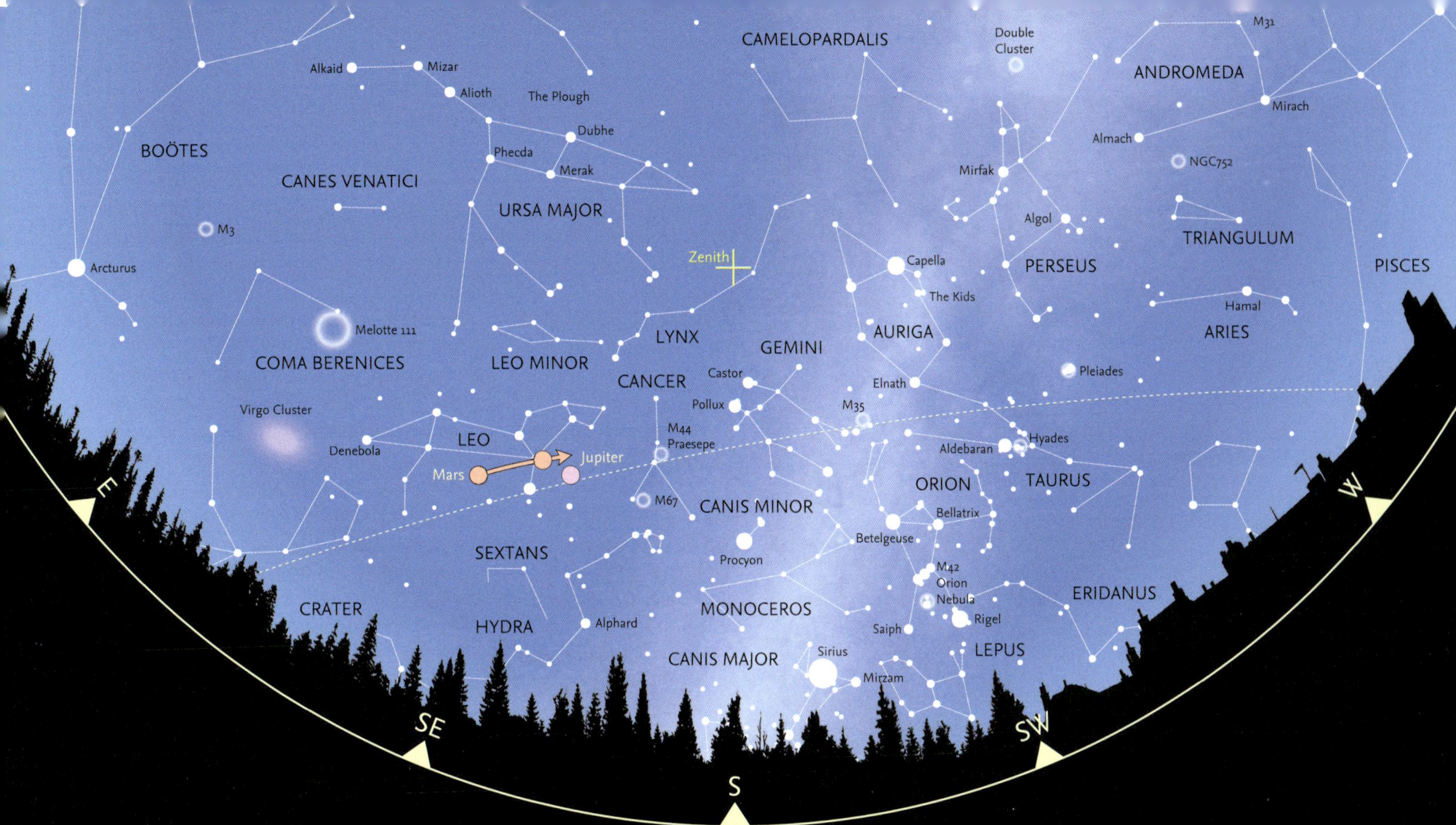

February – Looking South

1 February	23:00 GMT
15 February	22:00 GMT
1 March	21:00 GMT

February – Looking South

Apart from ***Orion***, the most prominent constellation visible this month is ***Gemini***; the heads of the twins are marked by the bright stars ***Castor*** and ***Pollux***, and the lines marking the bodies run southwest towards Orion. Castor (α Geminorum), the fainter star (mag. 1.9), is closer to the North Celestial Pole. Pollux (β Geminorum) is the brighter of the two (mag. 1.2) and it is farther away from the Pole. Pollux lies in the ecliptic and is occasionally occulted by the Moon. Castor is remarkable because it is actually a multiple system, consisting of no fewer than six individual stars arranged in three binary (paired) systems.

Orion's Belt points down to the southeast towards ***Sirius***, the brightest star in the sky (at mag. -1.4) in the constellation of ***Canis Major***, the whole of which is now clear of the southern horizon. Forming an equilateral triangle with ***Betelgeuse*** in Orion and Sirius in Canis Major is ***Procyon***, the brightest star in the small constellation of ***Canis Minor***. Between Canis Major and Canis Minor is the faint constellation of ***Monoceros***, which straddles the Milky Way. Although faint, there are many clusters in this area. Directly east of Procyon is the highly distinctive asterism of six stars that form the 'head' of ***Hydra***, the largest of all 88 constellations, and which trails such a long way across the sky that it is only around midnight in mid-March that the whole constellation becomes visible.

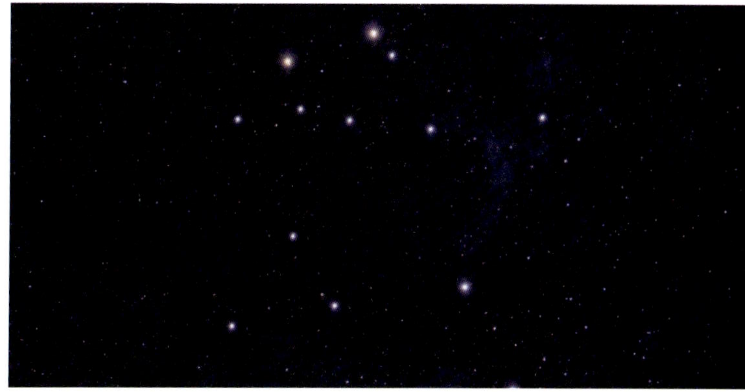

The constellation Gemini, with the heads of the twins marked by the stars Castor and Pollux.

The Moon's phases for February 2027

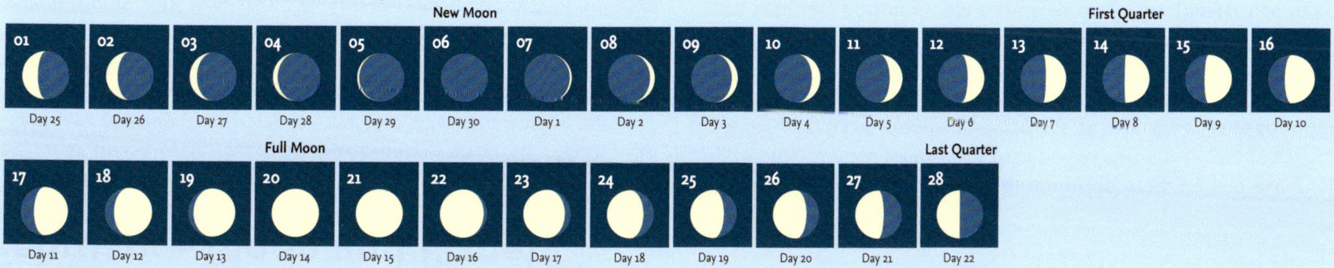

February – Moon and Planets

The Moon

The lunar phase cycle resets on 6 February with a New Moon. On 8 February ***Mercury*** (mag. 0.2) lies 0.4°N of the slim waxing crescent Moon. The First Quarter Moon arrives 14 February and the ***Pleiades*** can be seen 1.2°S of the Moon. The waxing gibbous Moon passes 4.4°S of ***Pollux*** on 18 February and two days later ***Jupiter*** (mag. -2.5) sits 1.2°N of the Full Moon, with ***Regulus*** 1.5°N. The waning gibbous Moon passes 2.9°S of ***Spica*** on 24 February and reaches Last Quarter on 28 February. On the same day the red supergiant ***Antares*** can be seen 0.4°N of the Moon, appearing low in the southeast a few hours before sunrise.

The Planets

Mercury is visible just after sunset at the beginning of the month at mag. -0.7, reaching eastern elongation on 3 February and switching to the early morning after 18 February, dimming to mag. 5.3 and eventually brightening to mag. 1.4 towards the end of February. It passes 0.4°N of the waxing crescent Moon on 8 February (at mag. 0.2). ***Venus*** continues to lead the Sun at dawn at mag. -4.3 and dims to mag. -4.1 at the end of the month. ***Mars*** rises several hours after sunset in ***Leo*** at mag. -0.9, it reaches opposition on 19 February at mag. -1.3, sits 3.8°N of ***Regulus*** on 22 February and ends the month at mag. -1.2. ***Jupiter*** lies in Leo with Mars and reaches opposition on 11 February at mag. -2.6. On 20 February Jupiter (mag. -2.5) lies 1.2°N of the Full Moon. ***Saturn*** moves through ***Cetus*** at the beginning of the month and arrives at ***Pisces*** from 23 February, visible in the early evening at mag. 0.8. ***Uranus*** sits in ***Taurus*** most of the night, at mag. 5.7. It ends retrograde motion on 8 February. ***Neptune*** is in Pisces at mag. 7.8 close to Saturn.

The path of the Sun and the planets along the ecliptic in February.

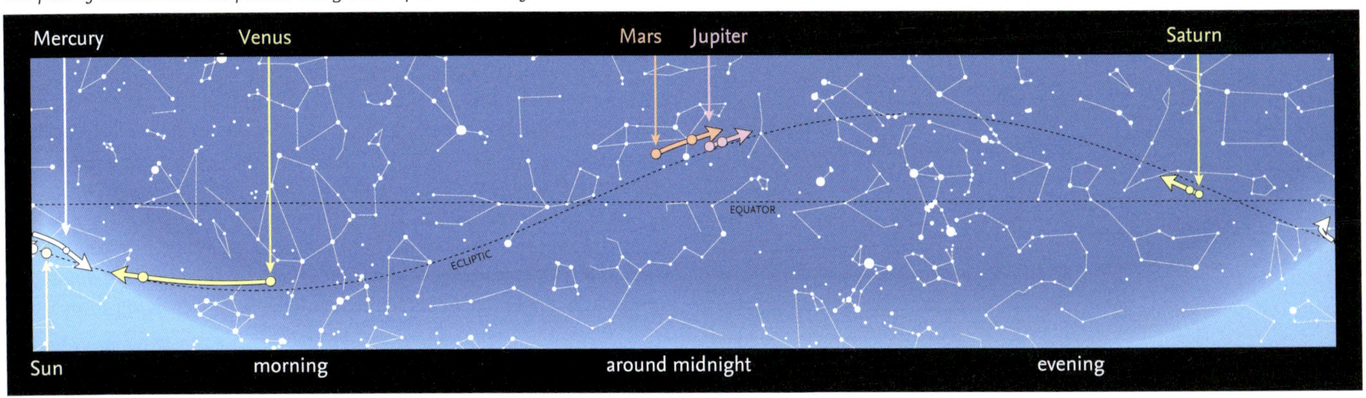

Calendar for February

03	05:11	Mercury at greatest elongation (18.3°E, mag. -0.5)
03	13:31	Moon at apogee = 406,189 km
06	15:56	New Moon
06	15:59	Annular solar eclipse (southern Chile, Argentina, Atlantic), partial eclipse (South America, Antarctica, west and south Africa)
08	03:36	Mercury (mag. 0.2) 0.4°N of the Moon
08	11:47	Uranus ends retrograde motion
11	00:21	Jupiter (mag. -2.6) at opposition
14	07:58	First Quarter Moon
14	16:13	Pleiades 1.2°S of the Moon
18	09:50	Pollux 4.4°N of the Moon
19	07:30	Moon at perigee = 361,015 km
19	15:44	Mars (mag. -1.3) at opposition
20	05:46	Jupiter (mag. -2.5) 1.2°N of the Moon
20	20:33	Regulus 1.5°N of the Moon
20	23:12	Penumbral lunar eclipse. Visible from the Americas, Africa, Asia, Russia, Antarctica and Europe
20	23:23	Full Moon
22	04:31	Mars (mag. -1.3) 3.8°N of Regulus
24	17:14	Spica 2.9°N of the Moon
28	05:16	Last Quarter Moon
28	06:09	Antares 0.4°N of the Moon

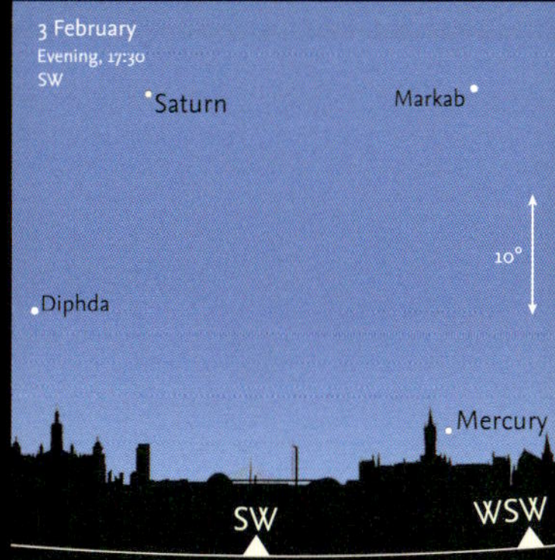

3 February • *Mercury at greatest eastern elongation.*

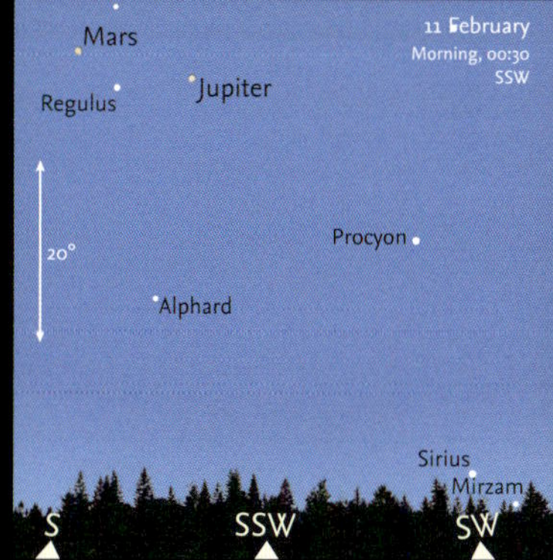

11 February • *Jupiter at opposition due south. Mars close by.*

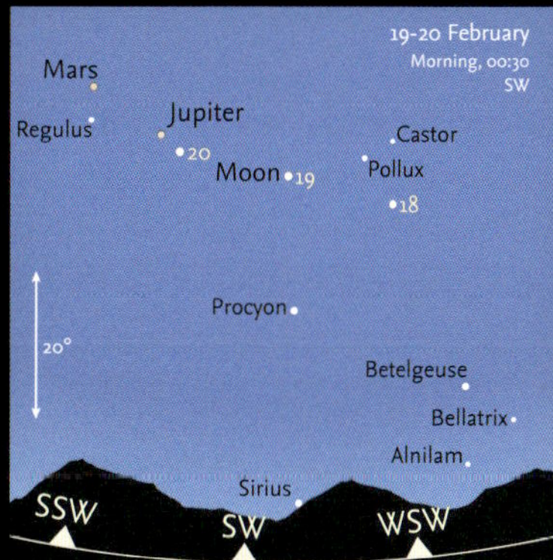

19–20 February • *Mars at opposition, the following day Jupiter sits next to the Full Moon.*

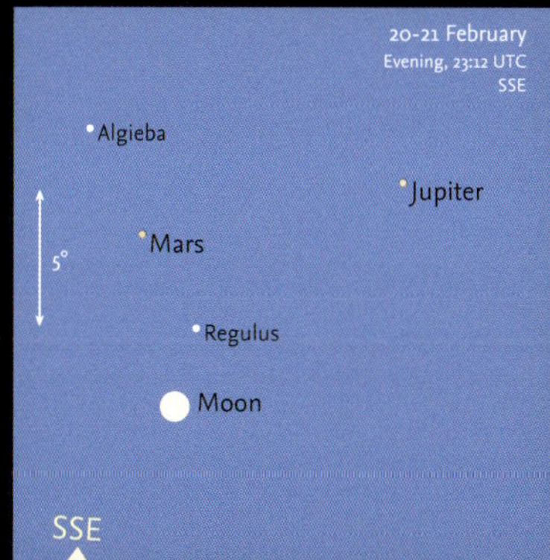

20–21 February • *Penumbral lunar eclipse, Mars and Jupiter nearby.*

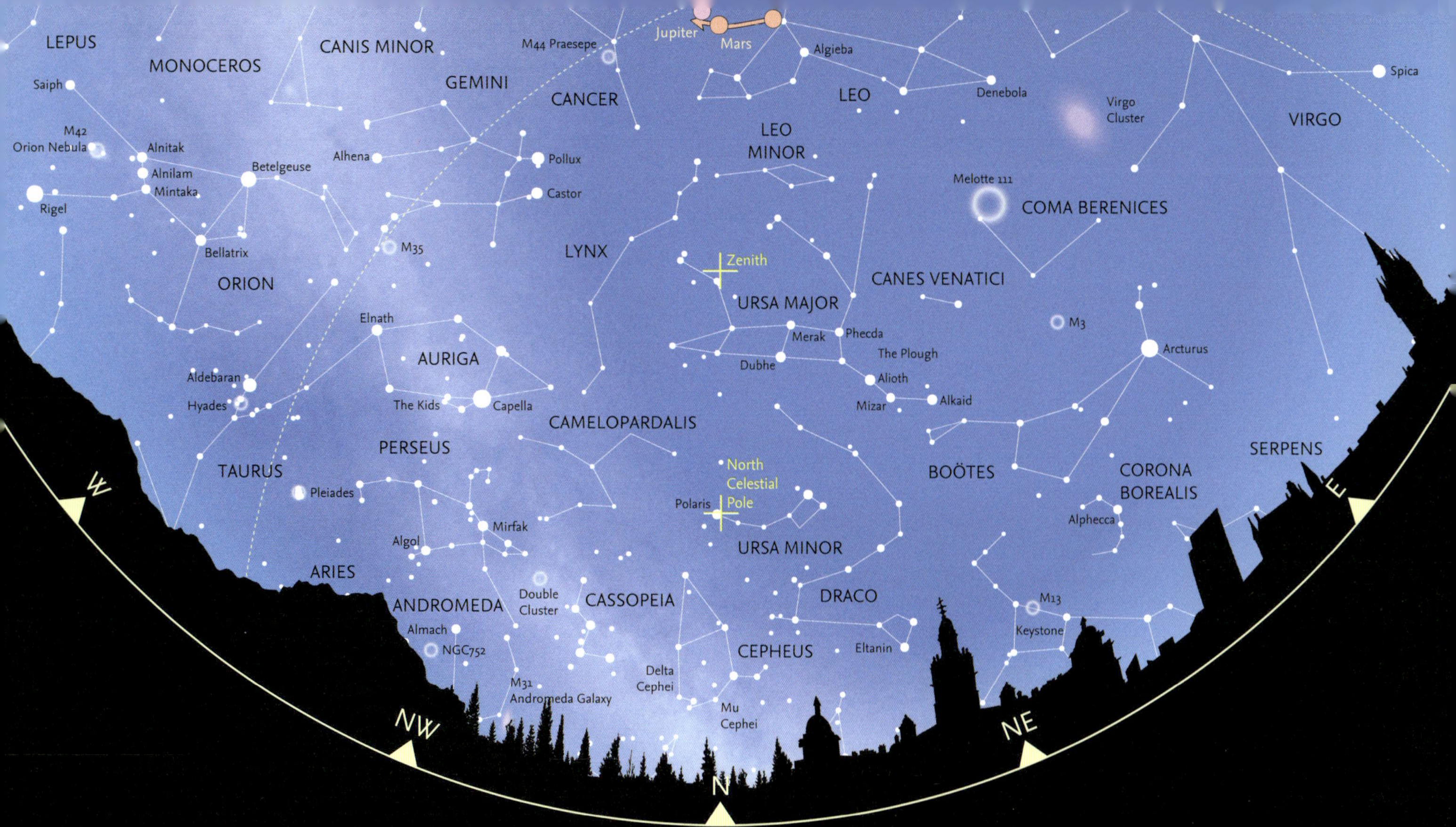

March – Looking North

1 March	23:00 GMT
15 March	22:00 GMT
1 April	22:00 BST (GMT + 1hr)

March – Looking North

The Sun crosses the celestial equator on 20 March, at the vernal equinox, when day and night are of almost equal length, and the northern season of spring is considered to begin. The hours of daylight change most rapidly around the equinoxes in March and September. It is also in March that British Summer Time (BST) begins (on Sunday 28 March) so the charts show the appearance at 23:00 for 1 March and 22:00 BST for 1 April. (In the rest of Europe, Daylight Saving Time is introduced on the same date.)

Early in the month, the constellation of ***Cepheus*** lies almost due north, with the distinctive 'W' of ***Cassiopeia*** to its west. Cepheus lies across the border of the Milky Way and is often described as like the gable end of a house or a church tower and steeple. One star stands out because of its deep red colour. This is ***μ Cephei***, also known as Herschel's Garnet Star and Erakis. It is a truly gigantic star, a red supergiant with a diameter 1,400 times that of the Sun, and if placed in the Solar System the surface of the star would extend beyond the orbit of Jupiter. ***Betelgeuse***, in ***Orion***, is also a red supergiant, however it is one-third of the size of μ Cephei.

Another famous and very important star in Cepheus is ***δ Cephei***, which is the prototype for the class of variable stars known as Cepheids. These giant stars show a regular variation in their luminosity, and there is a direct relationship between the period of the changes in magnitude and the stars' actual luminosity. By comparing its known luminosity with its apparent brightness, the star's distance can be determined. Once the distances to the closest Cepheid variables had been established, examples in more distant galaxies provided information about the scale of the universe. Cepheid variables are the first major 'rung' on the cosmic distance ladder.

Below Cepheus to the east (to the right), it may be possible to catch a glimpse of ***Deneb*** (α Cygni), just above the horizon. Slightly farther round towards the northeast, ***Vega*** (α Lyrae) is marginally higher in the sky. From southern Britain, Deneb is just far enough north to be circumpolar, rising higher in the early hours of the morning.

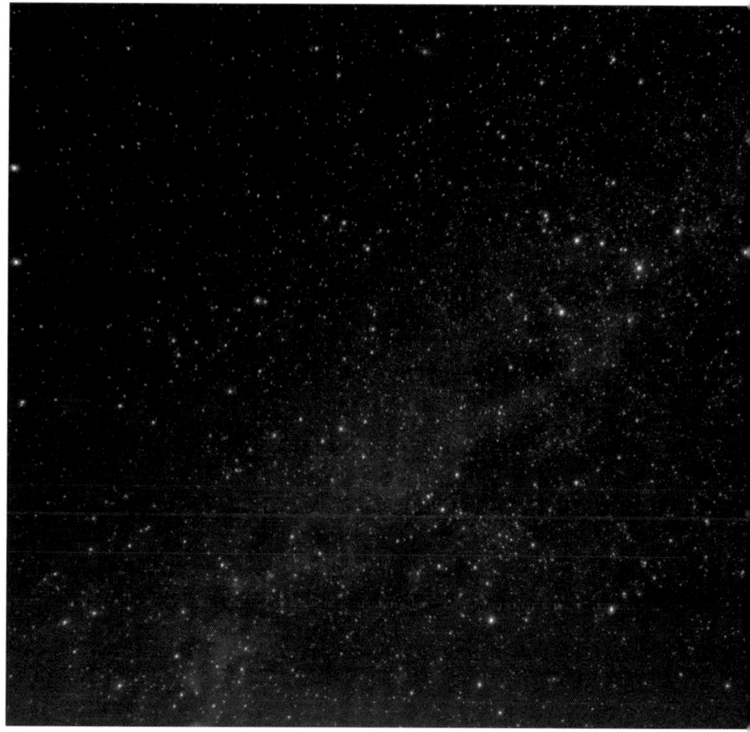

Cassiopeia, recognizable as a 'W', seen across the band of the Milky Way.

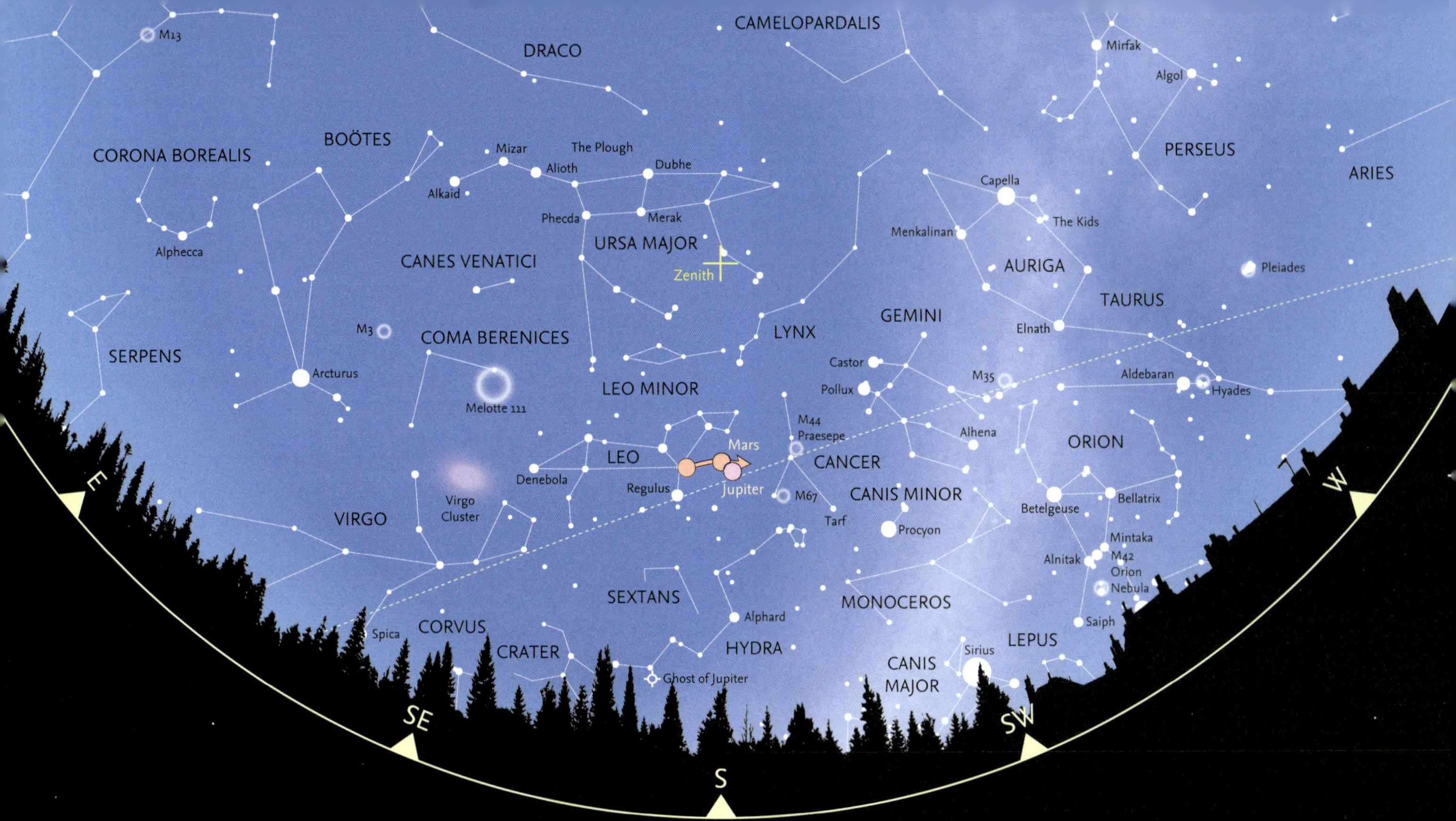

March – Looking South

1 March	23:00 GMT
15 March	22:00 GMT
1 April	22:00 BST (GMT + 1hr)

March – Looking South

Lying between the constellations of ***Gemini*** in the west and ***Leo*** in the east, and fairly high in the sky above the head of ***Hydra***, is the zodiacal constellation of ***Cancer***, the crab. Although a relatively faint constellation, its brightest star is β Cancri (also called ***Tarf***), an orange giant around 50 times the size of the Sun. Near the centre of the constellation lies an open cluster, M44 or ***Praesepe*** (the Manger but also known as the Beehive). On a clear night this cluster of around 1,000 stars, known since antiquity, is just a hazy spot to the naked eye, but appears in binoculars as a group of dozens of individual stars.

Also prominent in March is the constellation of Leo, with the 'backwards question mark' (or the Sickle) of bright stars forming the head of the mythological lion. ***Regulus*** (α Leonis) – the 'dot' of the question mark or the 'handle' of the sickle and the brightest star in Leo – lies very close to the ecliptic and is one of the few first-magnitude stars that may be occulted by the Moon. ***Denebola*** (β Leonis), a variable star, marks the tail of the lion.

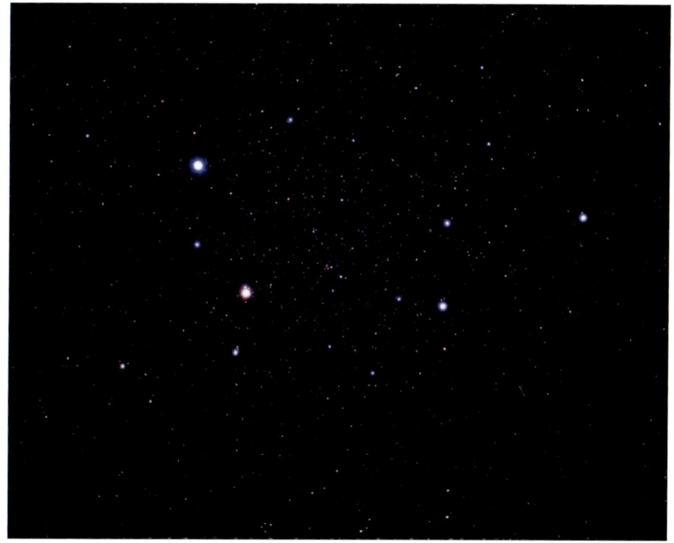

Leo the lion, with its 'backwards question mark'.

The Moon's phases for March 2027

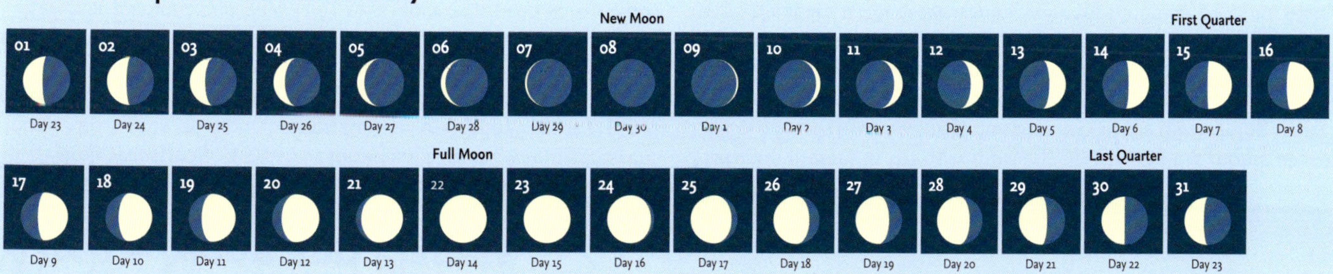

March – Moon and Planets

The Moon

On 4 March ***Venus*** (mag. -4.1) can be found 2.1°N of the waning crescent Moon – best seen an hour before sunrise the following day. On 6 March the thin crescent Moon passes 1.2°S of ***Mercury*** (mag. 0.6), also at dawn. The New Moon is on 8 March and five days later the ***Pleiades*** star cluster is placed 1.1°S of the Moon. The First Quarter Moon on 15 March is followed two days later by a close approach with ***Pollux***, sitting 4.5°N of the waxing gibbous Moon. On 19 March ***Jupiter*** (mag. -2.4) lies 0.9°N of the bright waxing gibbous Moon, with ***Mars*** (mag. -0.7) 4.3°N of the Moon. The following day the Moon appears 1.5°S of ***Regulus*** and it reaches a Full Moon on 22 March. ***Spica*** is placed 2.8°N of the Moon on 24 March, ***Antares*** sits closer at 0.3°N of the waning gibbous Moon on 27 March, above the horizon after midnight. The Moon reaches Last Quarter phase on 30 March.

The Planets

Mercury is present in the early morning sky starting the month at mag. 1.2. On 6 March Mercury (mag. 0.6) sits 1.2°N of the thin waning crescent Moon. It climbs to western elongation on 17 March and brightens to mag. -0.1 at the end of the month. ***Venus*** continues to lead the Sun at dawn at mag. -4.1, becoming gradually fainter during the month at mag. -3.9. It lies 2.1°N of the waning crescent Moon on 4 March. ***Mars*** is visible after sunset in ***Leo*** at mag. -1.2, dimming to mag. -0.4 at the end of March. On 19 March it passes 4.3°N of the waxing gibbous Moon at mag. -0.7. ***Jupiter*** lies in Leo with Mars and enters ***Cancer*** from 17 March, lying 0.9°N of the waxing gibbous Moon on 19 March and on the last day of March Mars (mag. -0.4) appears 4.3°N of Jupiter (mag. -2.3). ***Saturn*** stays in ***Pisces***, visible in the early evening at mag. 0.8 and brightening to 0.7. ***Uranus*** is visible in ***Taurus*** until midnight, at mag. 5.7 to 5.8. ***Neptune*** is in Pisces in the early evening at mag. 7.8.

The path of the Sun and the planets along the ecliptic in March.

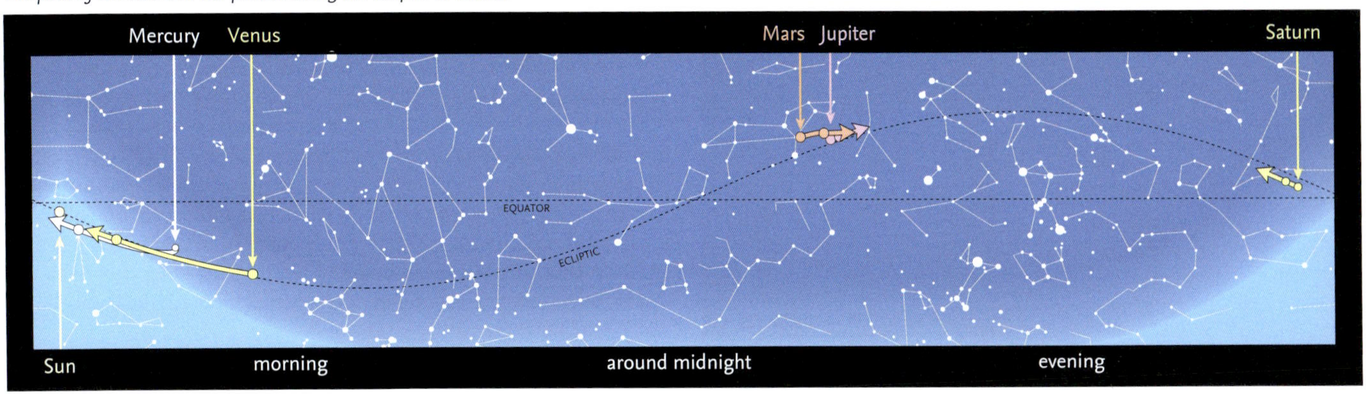

Calendar for March

03	05:41	Moon at apogee = 405,221 km
04	18:59	Venus (mag. -4.1) 2.1°N of the Moon
06	05:28	Mercury (mag. 0.6) 1.2°N of the Moon
08	09:29	New Moon
13	21:43	Pleiades 1.1°S of the Moon
15	16:25	First Quarter Moon
17	04:20	Mercury at greatest elongation (27.6°W, mag. 0.2)
17	17:40	Pollux 4.5°N of the Moon
19	04:31	Moon at perigee = 366,440 km
19	10:05	Jupiter (mag. -2.4) 0.9°N of the Moon
19	18:33	Mars (mag. -0.7) 4.3°N of the Moon
20	06:13	Regulus 1.5°N of the Moon
20	20:25	Vernal Equinox
22	10:44	Full Moon
24	03:12	Spica 2.8°N of the Moon
27	14:36	Antares 0.3°N of the Moon
30	00:54	Last Quarter Moon
31	01:33	Moon at apogee = 404,331 km

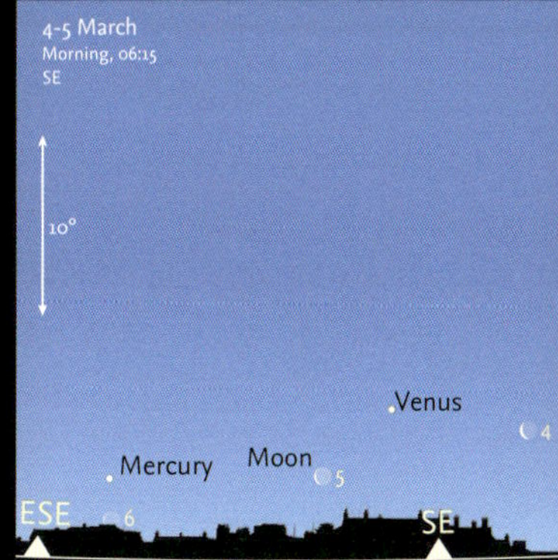

4–5 March • *Venus close to the waning crescent Moon, Mecury low in the sky.*

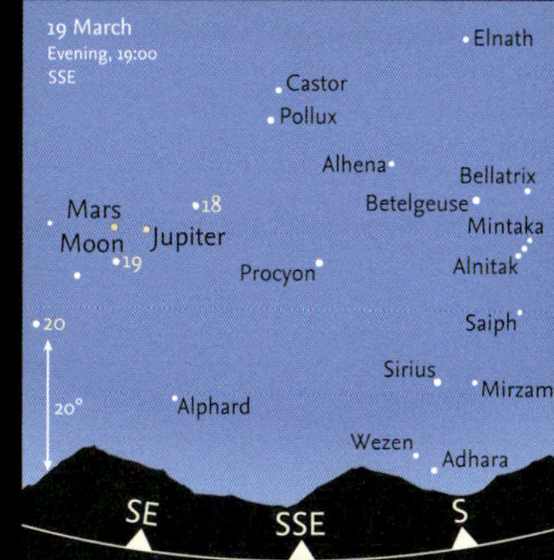

19 March • *Waxing gibbous Moon near to Mars and Jupiter.*

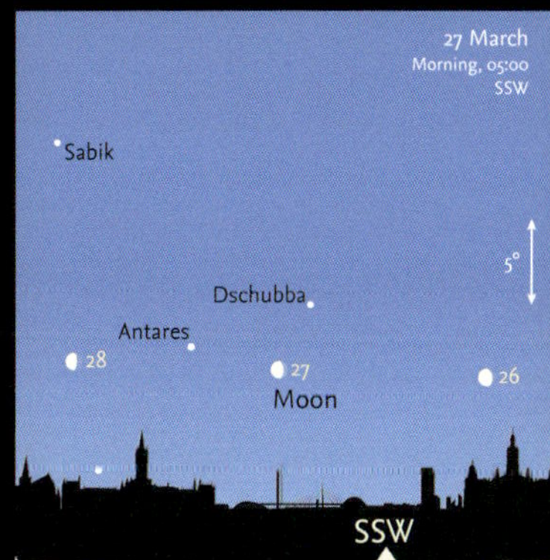

27 March • *Red star Antares close to the waning gibbous Moon.*

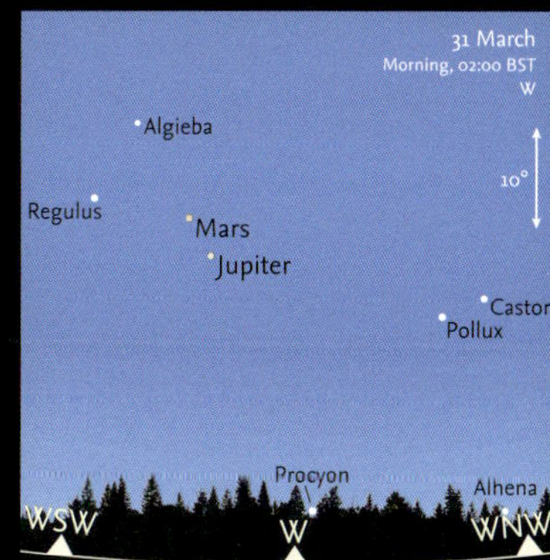

31 March • *Mars and Jupiter with Regulus.*

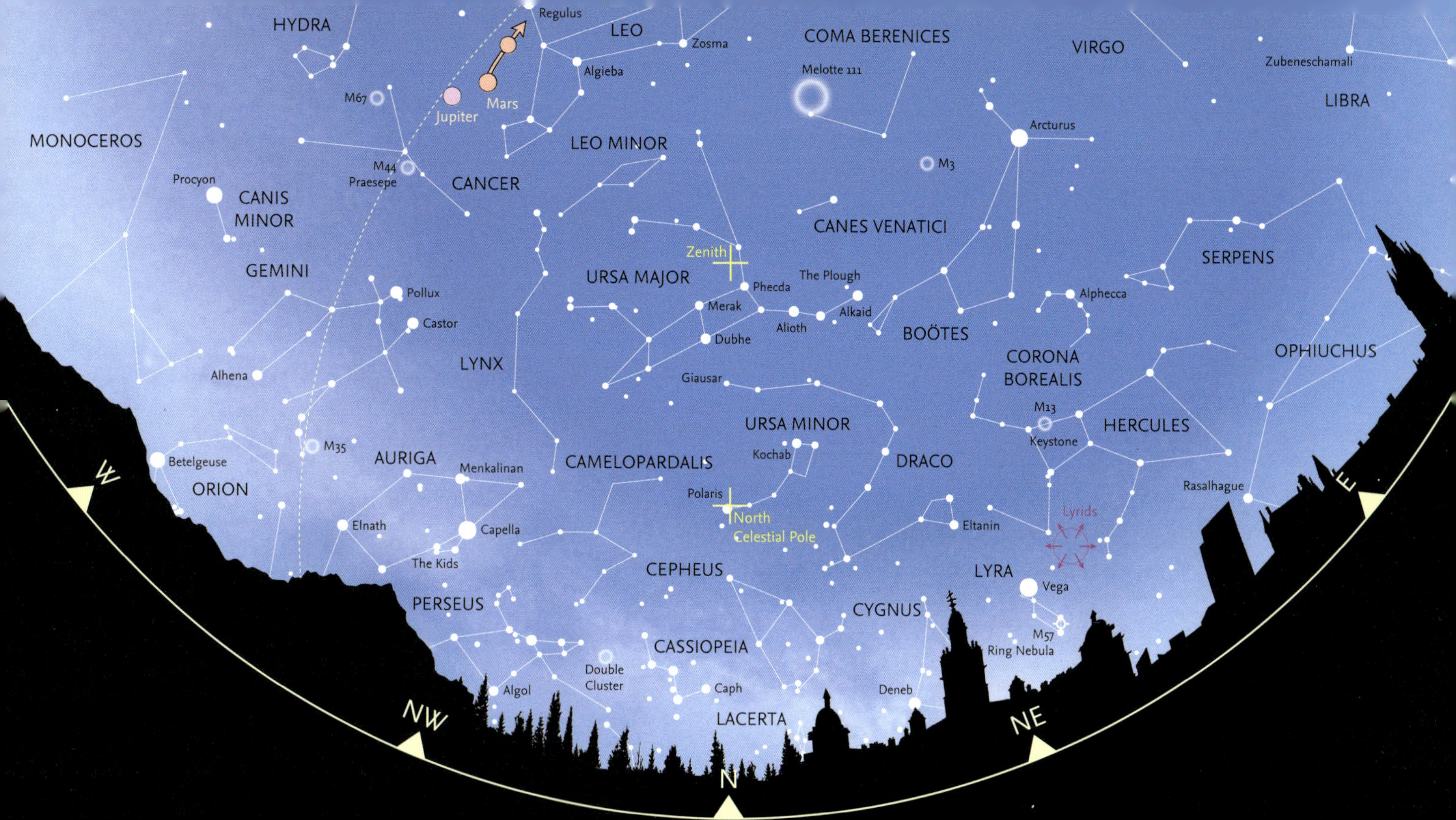

April – Looking North

1 April	00:00 BST (GMT + 1hr)
15 April	23:00 BST (GMT + 1hr)
1 May	22:00 BST (GMT + 1hr)

April – Looking North

Cygnus and the brighter regions of the Milky Way are now becoming visible, running almost parallel with the horizon in the early part of the night. Rising in the northeast is the small constellation of ***Lyra*** and the distinctive Keystone of ***Hercules*** above it. This asterism is very useful for locating the globular cluster ***M13***, which lies on one side of the quadrilateral. Some of the stars are 12 to 13 billion years old. The winding constellation of ***Draco*** weaves its way from the four stars that mark its 'head', on the border with Hercules, to end at ***Giausar*** (λ Draconis) between ***Polaris*** (α Ursae Minoris) and the Pointers, ***Dubhe*** and ***Merak*** (α and β Ursae Majoris, respectively). ***Ursa Major*** is 'upside down' high overhead, near the zenith. The constellation of ***Gemini*** stands almost vertically in the west. ***Auriga*** is still clearly seen in the northwest with its prominent star ***Capella***, but, by the end of the month, the southern portion of ***Perseus*** is starting to dip below the northern horizon. The very faint constellation of ***Camelopardalis*** lies in the northwest between Polaris and the constellations of Auriga and Perseus.

Meteors

There is one moderate meteor shower this month, the ***Lyrids*** (often known as the ***April Lyrids***). This year the shower begins on 16 April and peaks on the night of 23 April, a few days after the Full Moon. The sky will be darkest just before the Moon rises at 21:47, however more meteors will be seen after midnight as the Earth will have turned towards the cloud of debris, and it will be easier to spot them in the sky. There is another stronger shower, the ***η-Aquariids***, which begins on 19 April and peaks in the early hours of 5–6 May.

The Lyrid meteor shower peaks in April.

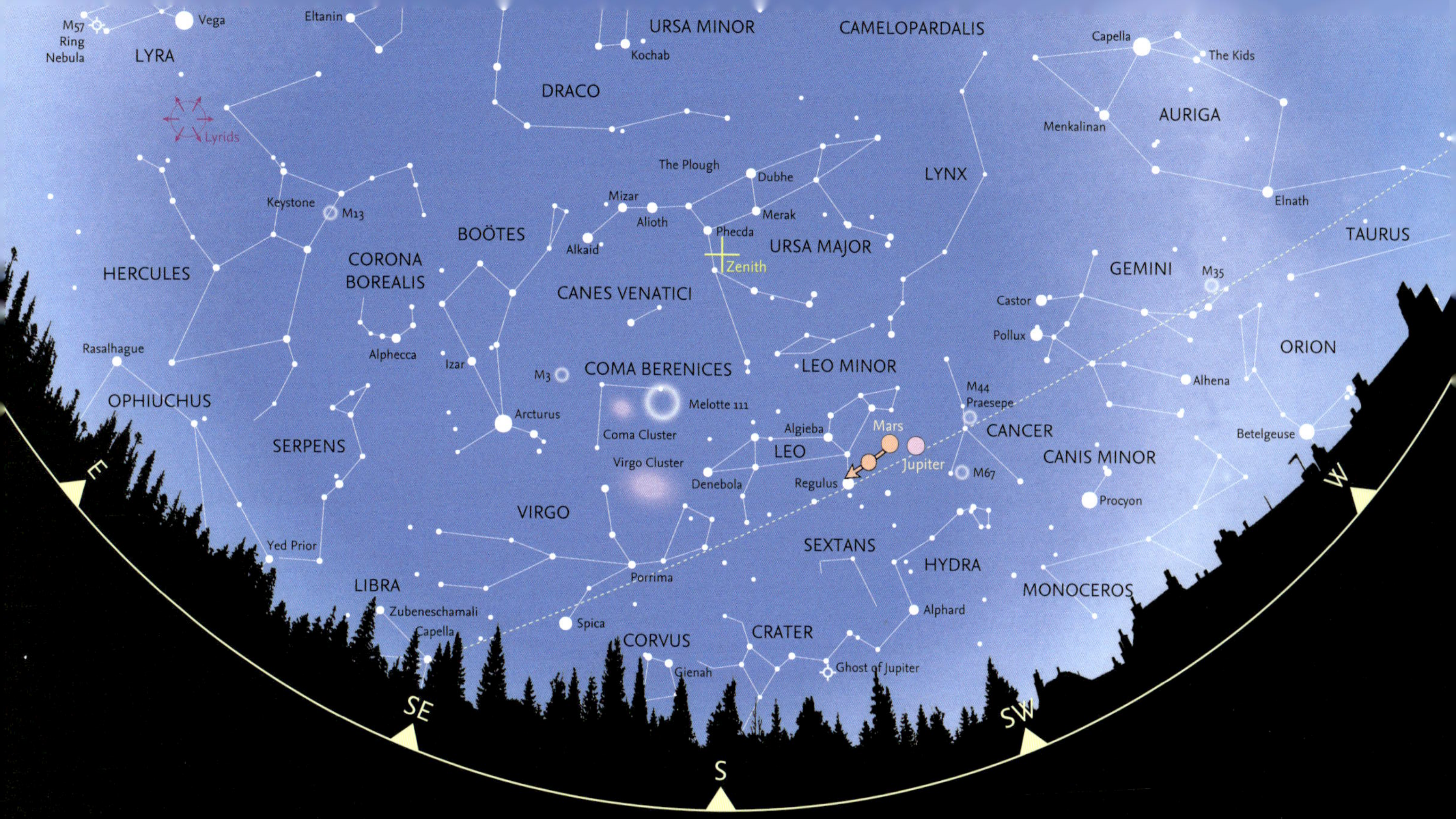

April – Looking South

1 April 00:00 BST (GMT + 1hr)
15 April 23:00 BST (GMT + 1hr)

April – Looking South

Leo is the most prominent constellation in the southern sky in April and looks vaguely like the creature after which it is named. ***Gemini***, with ***Castor*** and ***Pollux***, remains clearly visible in the west, and ***Cancer*** lies between the two constellations. To the east of Leo, the whole of ***Virgo***, with ***Spica*** (α Virginis) its brightest star, is well clear of the horizon. Below Leo and Virgo, the complete length of ***Hydra*** is visible, running beneath both constellations, with ***Alphard*** (α Hydrae) halfway between ***Regulus*** and the southwestern horizon. Farther east, the two small constellations of ***Crater*** and the rather brighter ***Corvus*** lie between Hydra and Virgo.

Boötes and ***Arcturus*** are prominent in the eastern sky, together with the circlet of ***Corona Borealis***, framed by Boötes and the neighbouring constellation of ***Hercules***. Between Leo and Boötes lies the constellation of ***Coma Berenices***, notable for being the location of the open cluster ***Melotte 111*** (see page 41) and the ***Coma Cluster*** of galaxies (Abell 1656). There are about 1,000 galaxies in this cluster, which is located near the North Galactic Pole, where we are looking out of the plane of the Galaxy and are thus able to see deep into space. Only about ten of the brightest galaxies in the Coma Cluster are visible with the largest amateur telescopes.

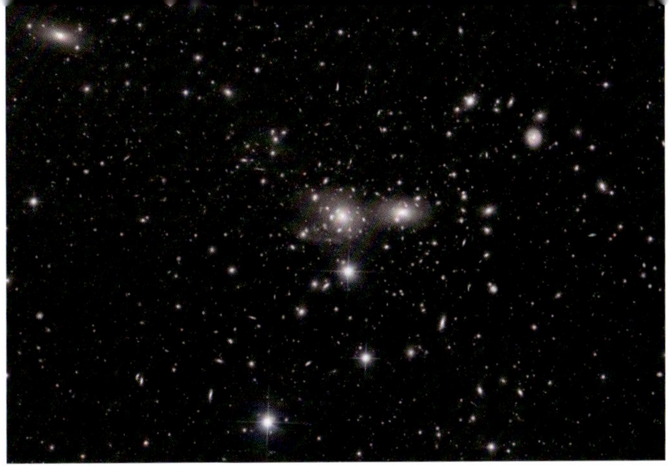

The Coma Cluster, a group of over 1,000 galaxies.

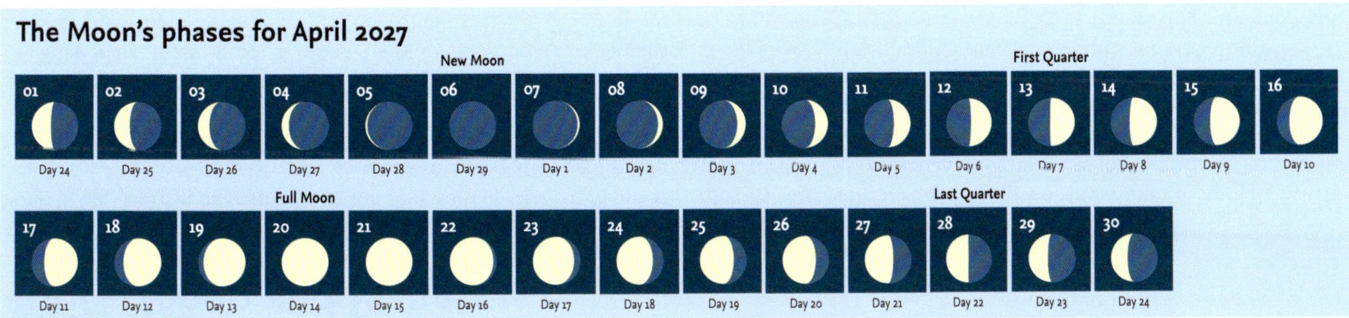

April – Moon and Planets

The Moon

Venus (mag. -3.9) can be seen 3.3°S of the slim waning crescent Moon at dawn on 4 April. Two days later the lunar phase cycle starts again with a New Moon. The waxing crescent Moon appears 0.9°N of the ***Pleiades*** on 10 April and on 13 April ***Pollux*** lies 4.7°N of the First Quarter Moon. On 15 April, ***Jupiter*** (mag. -2.2) sits 1.0°N of the waxing gibbous Moon, the day after ***Mars*** (mag. -0.02) is found 3.3°N of the Moon and ***Regulus*** 1.6°N of the Moon. On 20 April the Full Moon passes 2.8°S of ***Spica***, rising later in the evening. ***Antares*** appears close to the waning gibbous Moon, only 0.1°N on 23 April, both above the horizon in the early hours of the following day. The Last Quarter Moon arises on 28 April.

The Planets

Mercury rises with the Sun for most of the month of April. It switches to the evening sky after 28 April, brightening from mag. -0.1 to -2.3. ***Venus*** appears at dawn before the sunrise, easy to see at mag. -3.9. Venus will appear 3.3°S of the waning crescent Moon on 4 April. ***Mars*** ends retrograde motion on 1 April and can be seen in ***Leo*** after sunset at mag. -0.4, dimming to mag. 0.2 as the month progresses. On 16 April Mars (mag. -0.02) lies 3.3°N of the waxing gibbous Moon. ***Jupiter*** lies in ***Cancer***, visible after sunset and dimming from mag. -2.3 to -2.1. It ends retrograde motion on 13 April. On 15 April Jupiter (mag. -2.2) will sit 1.0°N of the waxing gibbous Moon. ***Saturn*** lies in ***Pisces***, setting with the Sun at mag. 0.7. ***Uranus*** is visible in ***Taurus*** in the early evening, at mag. 5.8. ***Neptune*** is in Pisces rising just before the Sun, at mag. 7.8.

The path of the Sun and the planets along the ecliptic in April.

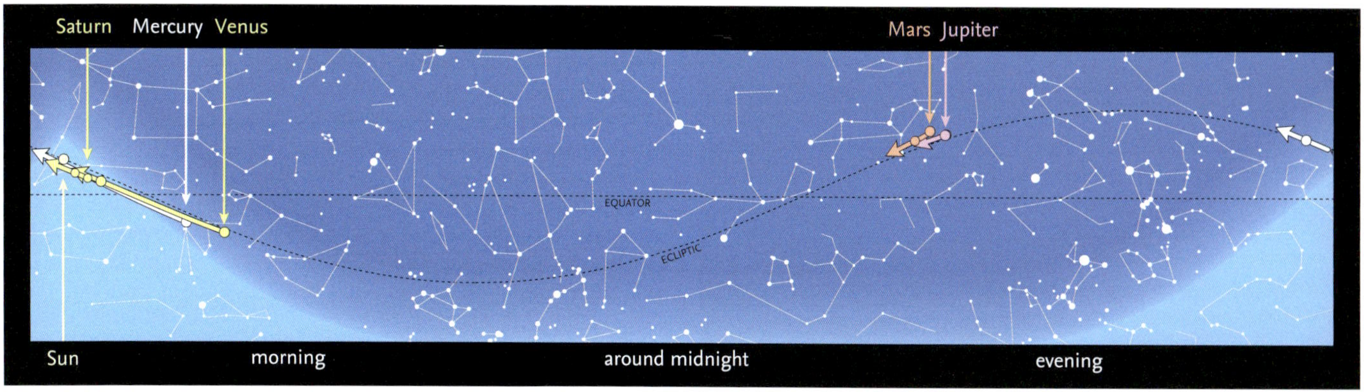

endar for April

15:07	Mars ends retrograde motion
06:49	Venus (mag. -3.9) 3.3°S of the Moon
23:51	New Moon
03:30	Pleiades 0.9°S of the Moon
03:24	Jupiter ends retrograde motion
22:57	First Quarter Moon
23:23	Pollux 4.7°N of the Moon
00:41	Moon at perigee = 370,001 km
15:01	Jupiter (mag. -2.2) 1.0°N of the Moon
01:00	Mars (mag. -0.02) 3.3°N of the Moon
13:21	Regulus 1.6°N of the Moon
12:09	Spica 2.8°N of the Moon
22:27	Full Moon
	Lyrid meteor shower maximum
23:16	Antares 0.1°N of the Moon
21:21	Moon at apogee = 404,170 km

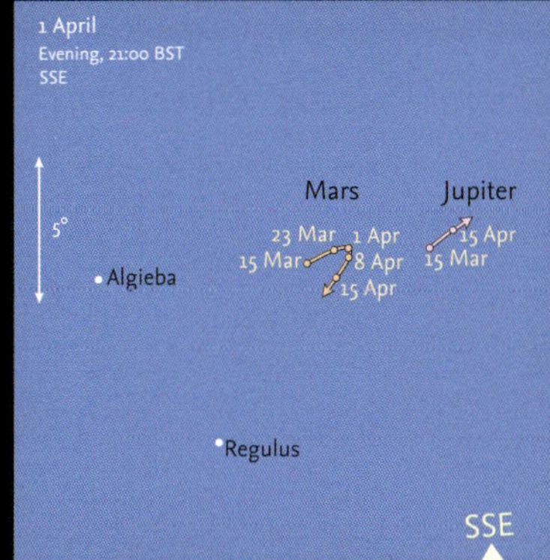

1 April • *Mars ends its retrograde motion in Leo.*

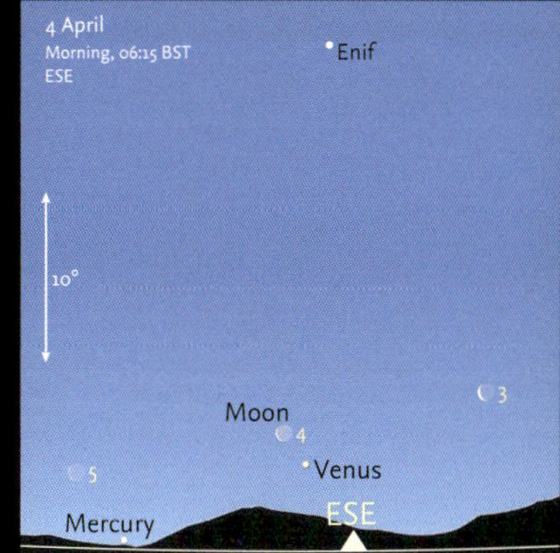

4 April • *Venus and the waning crescent Moon. Mecury skimming the horizon.*

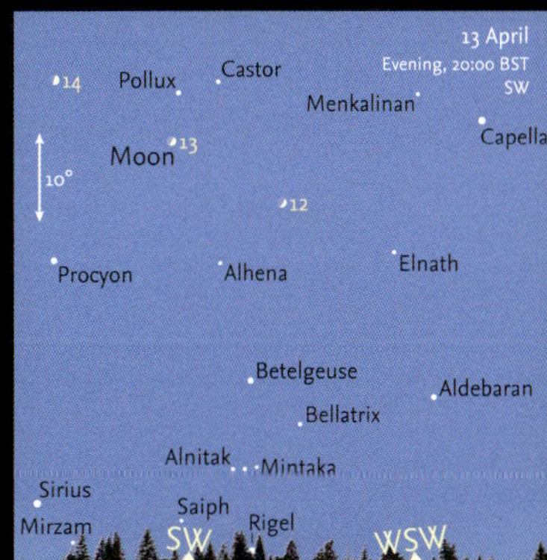

13 April • *First Quarter Moon close to Pollux in Gemini.*

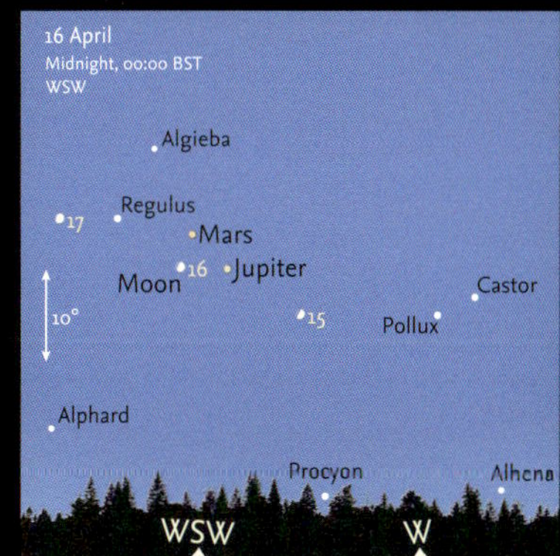

16 April • *Mars and Jupiter with Regulus and the waxing gibbous Moon.*

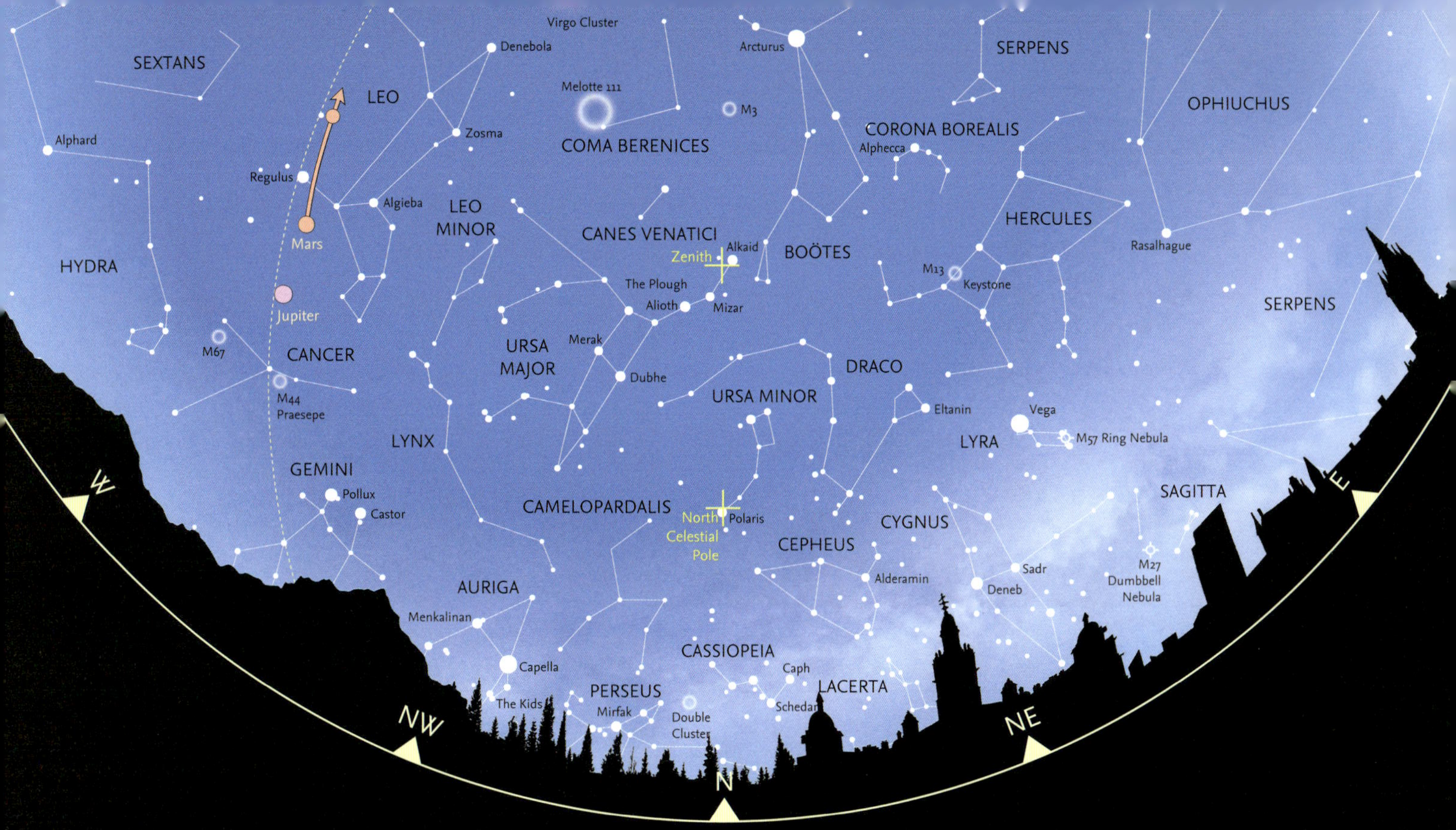

May – Looking North

1 May	00:00 BST (GMT + 1hr)
15 May	23:00 BST (GMT + 1hr)
1 June	22:00 BST (GMT + 1hr)

May – Looking North

Cassiopeia is now low over the northern horizon and, to its west, the southern portions of both ***Perseus*** and ***Auriga*** are becoming difficult to observe, although the ***Double Cluster***, between Perseus and Cassiopeia is still clearly visible. Each cluster is composed of more than 300 supergiants many times hotter than the Sun; some of these stars are nearing the end of their lives and have evolved into red supergiants. The ***Andromeda Galaxy*** is now too low to be readily visible. The constellations of ***Lyra***, ***Cepheus***, ***Ursa Minor*** and the whole of ***Draco*** are well placed in the sky. ***Gemini***, with ***Castor*** and ***Pollux***, is sinking towards the western horizon. ***Capella*** (α Aurigae) and the three stars in the asterism The Kids are still clear of the horizon.

In the east, two of the stars of the Summer Triangle, ***Vega*** (α Lyrae) and ***Deneb*** (α Cygni), are clearly visible, and the third star, ***Altair*** in ***Aquila***, is beginning to climb above the horizon. The whole of ***Cygnus*** is now visible. The sprawling constellation of ***Hercules*** is high in the east and the brightest globular cluster in the northern hemisphere, **M13**, is visible to the naked eye on the western side of the asterism known as the Keystone.

Three faint constellations may be identified before the lighter nights of summer make them difficult objects. Below Cepheus, in the northeastern sky is the zig-zag constellation of ***Lacerta***, while to the west, above Perseus and Auriga is ***Camelopardalis*** and, farther west, the line of faint stars forming ***Lynx***.

Later in the night (and in the month) the westernmost stars of ***Pegasus*** begin to come into view. High overhead, ***Alkaid*** (η Ursae Majoris), the last star in the 'tail' of the Great Bear, is close to the zenith, while the main body of the constellation has swung round into the western sky.

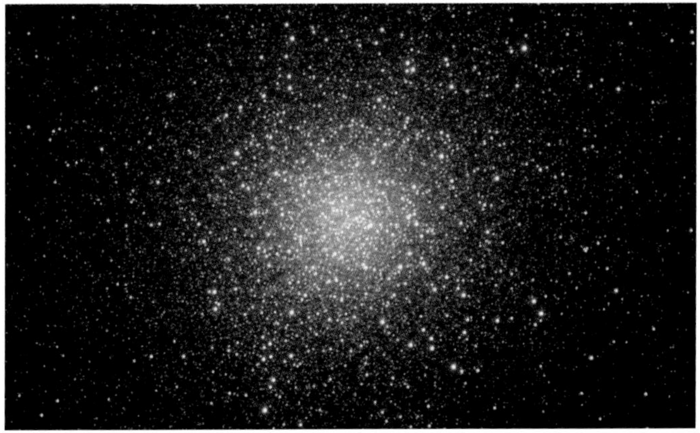

M13, the Great Globular Cluster, in Hercules.

Meteors

The ***η-Aquariids*** are one of the two meteor showers associated with Comet 1P/Halley, Halley's Comet, last seen in 1986 and not due until 2061. The other shower arising from the cometary debris is the ***Orionids***, in October. The h-Aquariids are not favourably placed for observers in the northern hemisphere, because the radiant is well below the horizon until shortly before dawn. However, meteors may still be seen in the eastern sky even when the radiant is below the horizon. There is a radiant map for the h-Aquariids on page 30.

The peak of the shower in the early hours of 6 May coincides with a New Moon, meaning optimum conditions for spotting faint meteors. Maximum hourly rate is about 40 per hour (around half that visible from the northern hemisphere), and a quarter of the meteors leave persistent trains, which can last for several seconds to minutes.

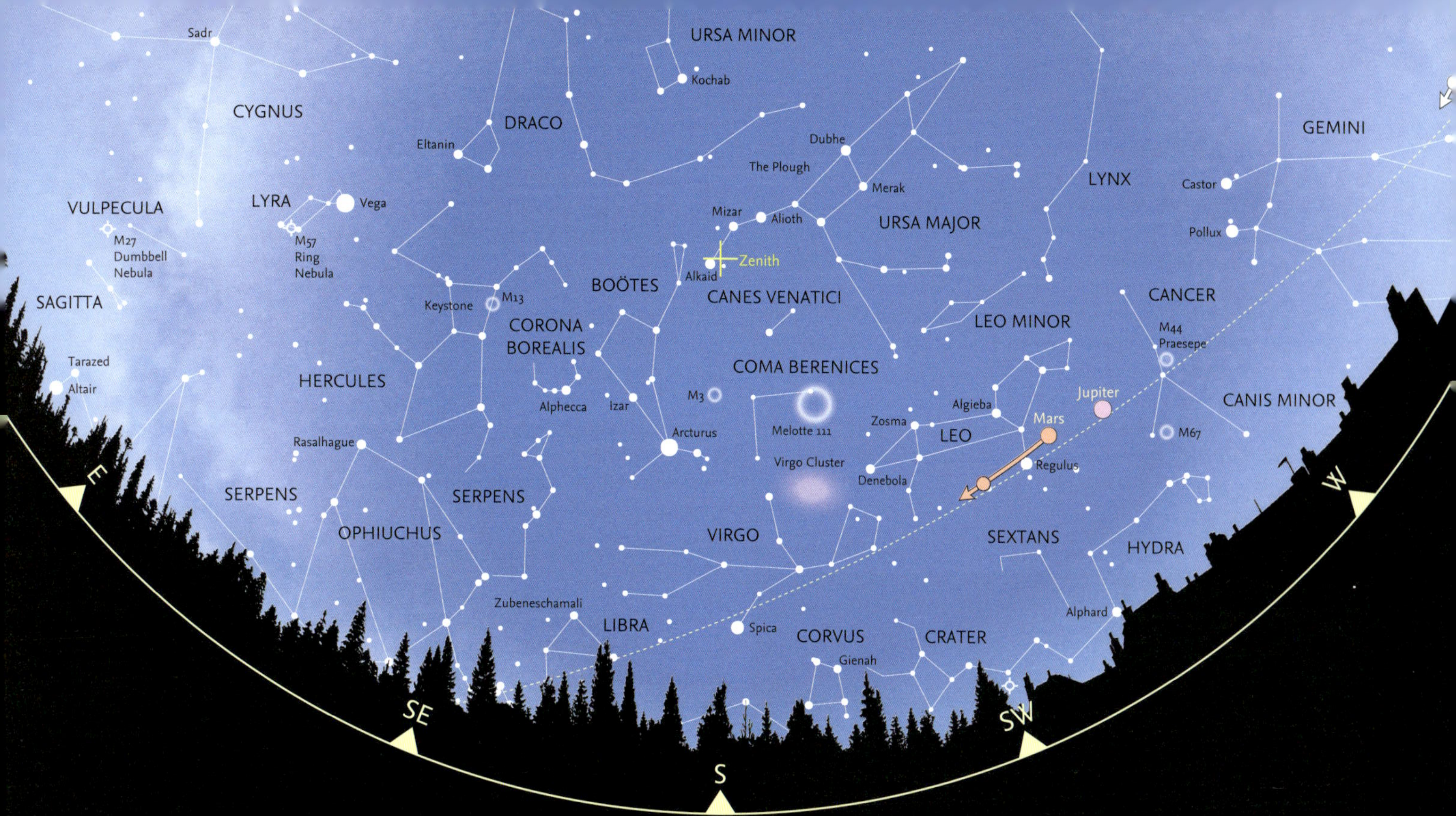

May – Looking South

1 May	00:00 BST (GMT + 1hr)
15 May	23:00 BST (GMT + 1hr)
1 June	22:00 BST (GMT + 1hr)

May – Looking South

Early in the night, the constellation of ***Virgo***, with ***Spica*** (α Virginis), lies due south, with ***Leo*** and both ***Regulus*** and ***Denebola*** (α and β Leonis, respectively) to its west still well clear of the horizon. Later in the night, the rather faint zodiacal constellation of ***Libra*** becomes visible and, to its east, the red-orange star ***Antares*** (α Scorpii) begins to climb up over the horizon.

Virgo contains the nearest large cluster of galaxies, which is the centre of the Local Supercluster, of which the Milky Way galaxy forms part. The Virgo Cluster contains some 2,000 galaxies, the brightest of which are visible in amateur telescopes, several of which are in the Messier catalogue. ***Arcturus*** in ***Boötes*** is high in the south, with the distinctive circlet of ***Corona Borealis*** clearly visible to its east. The brightest star (α Coronae Borealis) is known as ***Alphecca***. The large constellation of ***Ophiuchus*** (which actually crosses the ecliptic and is thus the 'thirteenth' zodiacal constellation) is climbing into the eastern sky. Before the constellation boundaries were formally adopted by the International Astronomical Union in 1930, the southern region of Ophiuchus was regarded as forming part of the constellation of Scorpius, which had been part of the zodiac since antiquity.

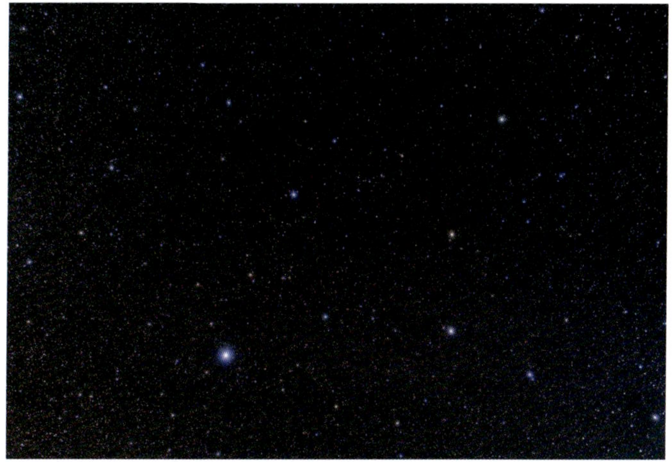

Virgo ('the maiden') and its brightest star, Spica.

The Moon's phases for May 2027

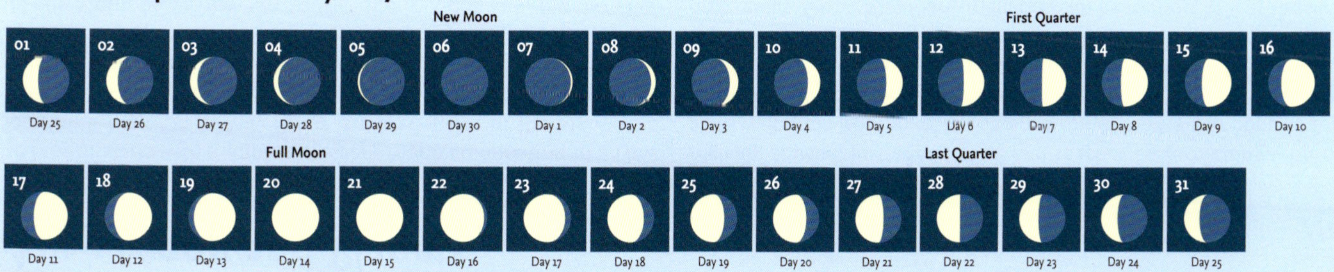

May – Moon and Planets

The Moon

The month starts with a waning crescent Moon culminating in a New Moon on 6 May. On 11 May ***Pollux*** is placed 4.9°N of the Moon, the following night ***Jupiter*** (mag. -2.0) appears 1.3°N of the Moon, high in the sky after sunset. The First Quarter Moon, on 13 May, is 1.9°S of ***Regulus*** and 3.3°S of ***Mars*** (mag. 0.4). ***Spica*** sits 2.9°N of the waxing gibbous Moon on 17 May. On 20 May the Moon is Full, the day after ***Antares*** and the Moon are less than a tenth of a degree apart a few hours before sunrise. The Moon reaches Last Quarter phase on 28 May.

The Planets

Mercury appears just after sunset and reaches eastern elongation on 28 May, it dims from mag. -2.1 to 0.7 through May. ***Venus*** pops up just above the horizon before sunrise at mag. -3.9. It approaches ***Saturn*** (mag. 0.7) on 7 May, passing 0.6°N of the ringed gas giant. ***Mars*** stays in ***Leo*** after sunset at mag. 0.3, dimming to mag. 0.7 as the month progresses. On 13 May Mars (mag. 0.4) can be found 3.3°N of the First Quarter Moon and the day after it will be 1.1°N of ***Regulus***. ***Jupiter*** lies in ***Cancer*** until 10 May, after which it moves into Leo. It is visible most of the night, dimming from mag. -2.1 to -1.9 at the end of May. Jupiter (mag. -2.0) lies 1.3°N of the waxing crescent Moon on 12 May. Saturn lies in ***Pisces***, rising just before the Sun at mag. 0.7. ***Uranus*** is visible in ***Taurus*** in the early evening, at mag. 5.8. ***Neptune*** is in Pisces rising just before the Sun, at mag. 7.8.

The path of the Sun and the planets along the ecliptic in May.

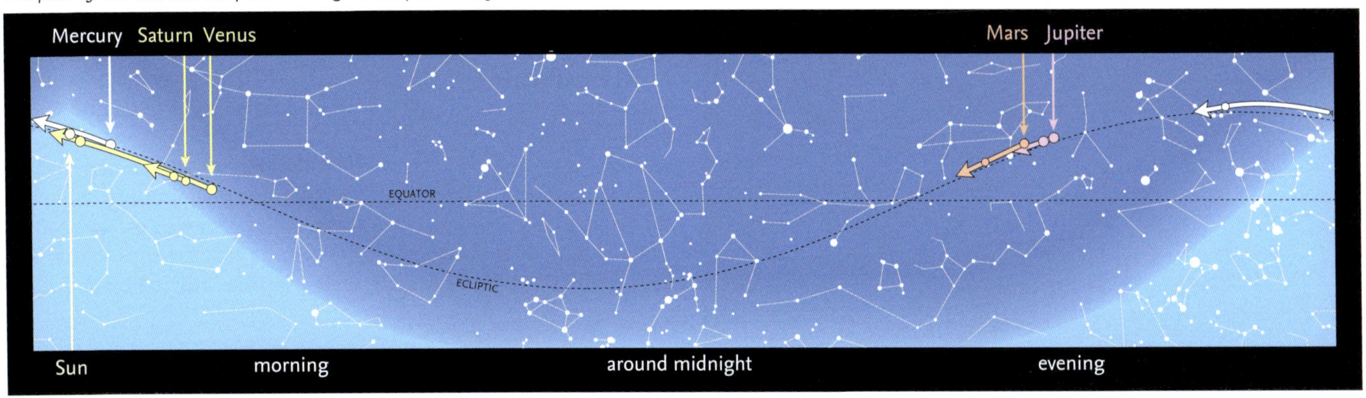

Calendar for May

05–06		Eta (η) Aquariid meteor shower maximum
06	10:58	New Moon
07	16:00	Venus (mag. -3.9) 0.6°N of Saturn (mag. 0.7)
09	20:07	Moon at perigee = 366,635 km
11	05:06	Pollux 4.9°N of the Moon
12	22:49	Jupiter (mag. -2.0) 1.3°N of the Moon
13	04:44	First Quarter Moon
13	18:57	Regulus 1.9°N of the Moon
13	19:34	Mars (mag. 0.4) 3.3°N of the Moon
14	10:58	Mars (mag. 0.4) 1.1°N of Regulus
17	19:16	Spica 2.9°N of the Moon
20	10:59	Full Moon
21	07:08	Antares 0.0°S of the Moon
25	15:13	Moon at apogee = 404,801 km
28	13:58	Last Quarter Moon
28	14:13	Mercury at greatest elongation (23.0°E, mag. 0.4)

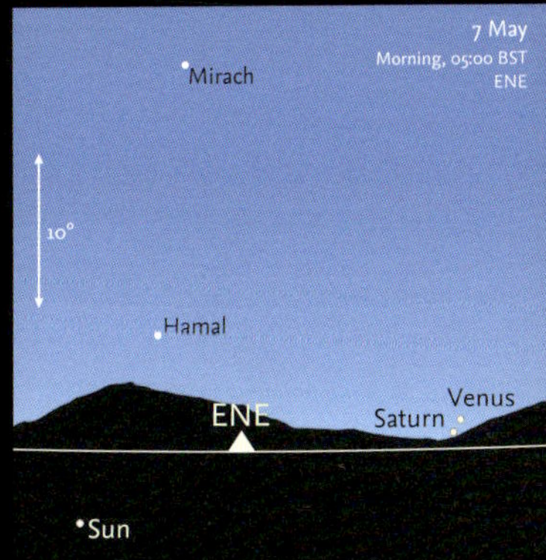

7 May • *Venus and Saturn low in the east.*

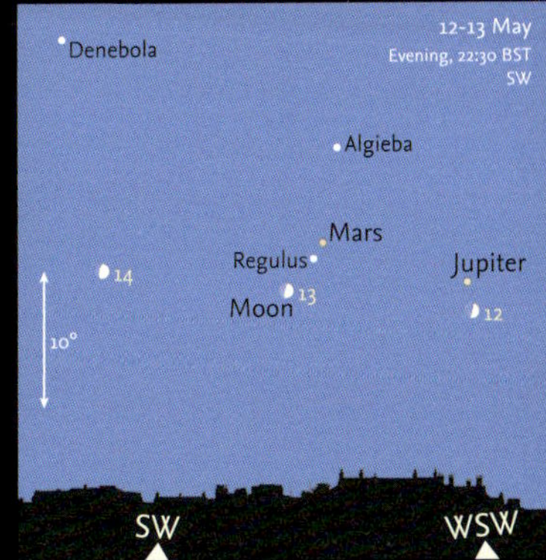

12–13 May • *Waxing crescent Moon swings by Jupiter, Mars and Regulus.*

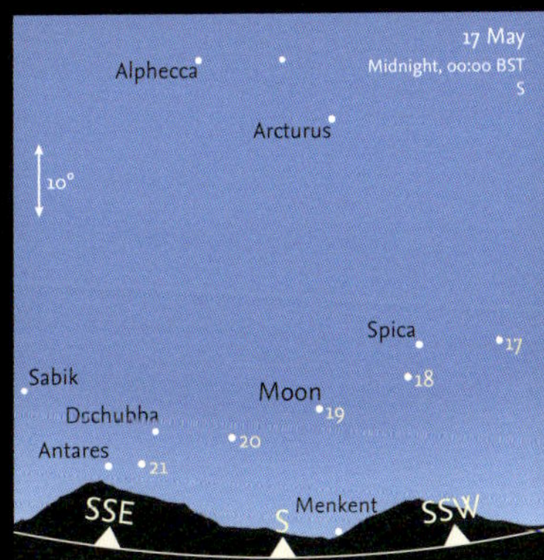

17 May • *Virgo's brightest star, Spica, close to the waxing gibbous Moon.*

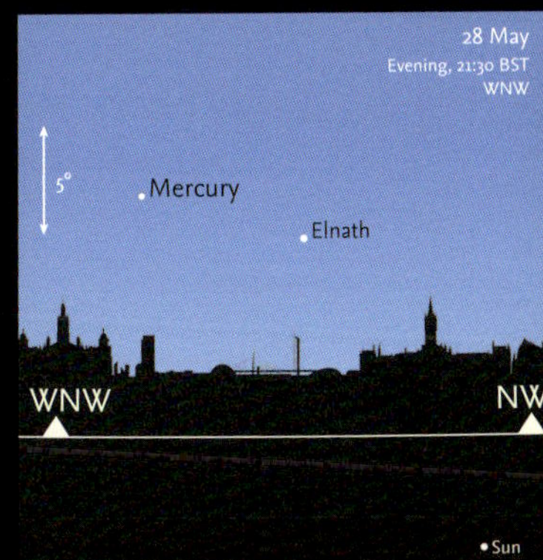

28 May • *Mercury at greatest eastern elongation.*

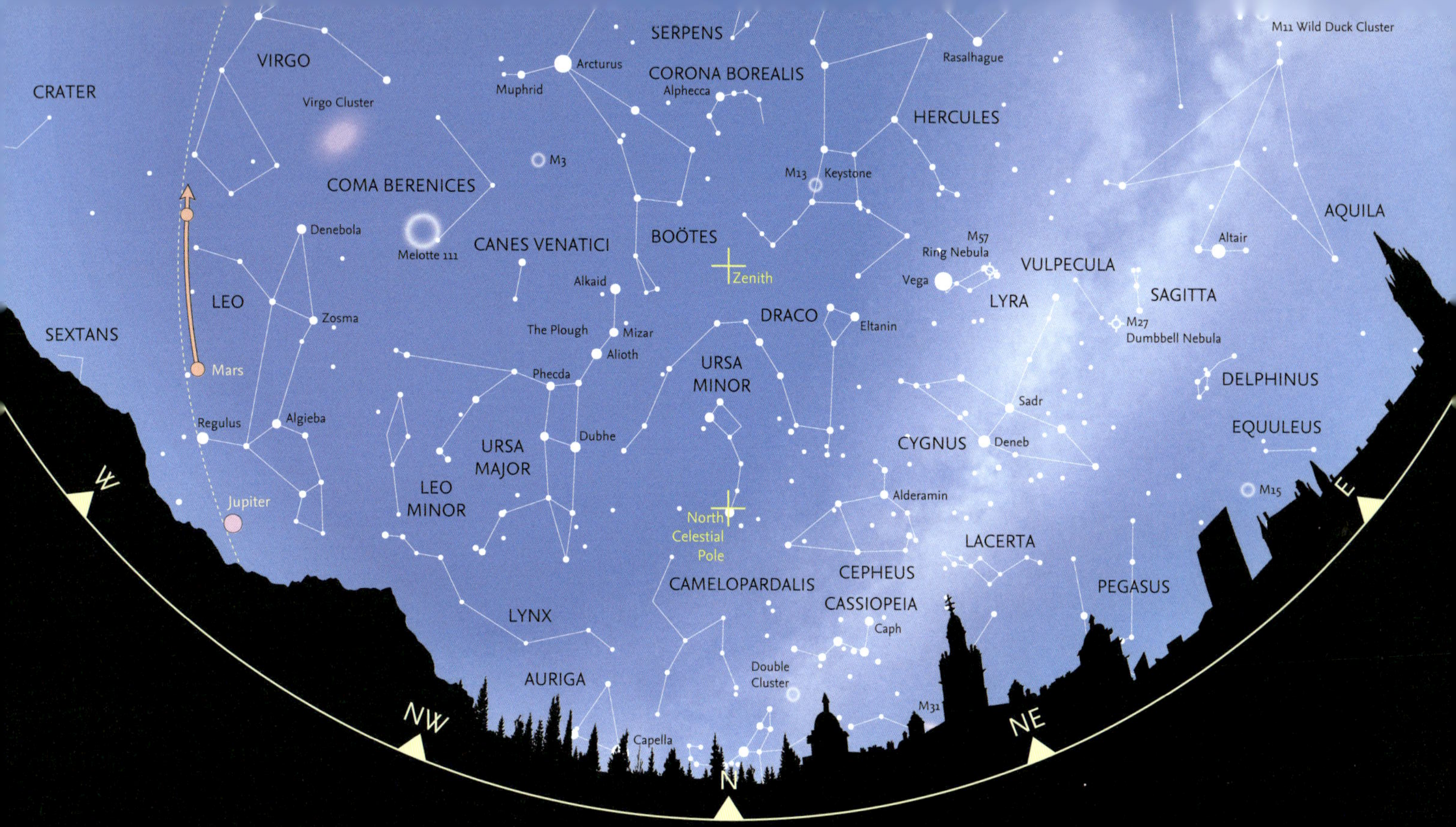

June – Looking North

1 June	00:00 BST (GMT + 1hr)
15 June	23:00 BST (GMT + 1hr)

June – Looking North

Around summer solstice (21 June), even in southern England and Ireland, a form of twilight persists throughout the night due to the Sun not setting as far below the horizon. Farther north, in Scotland, the sky remains so light that most of the fainter stars and constellations are invisible. There, even brighter stars such as the seven stars making up the well-known asterism known as the Plough in ***Ursa Major*** may be difficult to detect except around local midnight, 00:00 UT (01:00 BST).

But there is one compensation during these light nights: northern observers may be lucky enough to witness a display of noctilucent ('night-shining') clouds. These are highly distinctive clouds shining with an electric-blue tint, observed in the sky in the direction of the North Pole. They are the highest clouds in the atmosphere, occurring at altitudes of 80–85 km, and they are composed of ice crystals. They are only visible during summer nights shortly after sunset and before sunrise, for up to six weeks on either side of the solstice, but the clouds themselves remain illuminated by sunlight. They act like high mirrors, reflecting sun rays when the Sun is below the horizon.

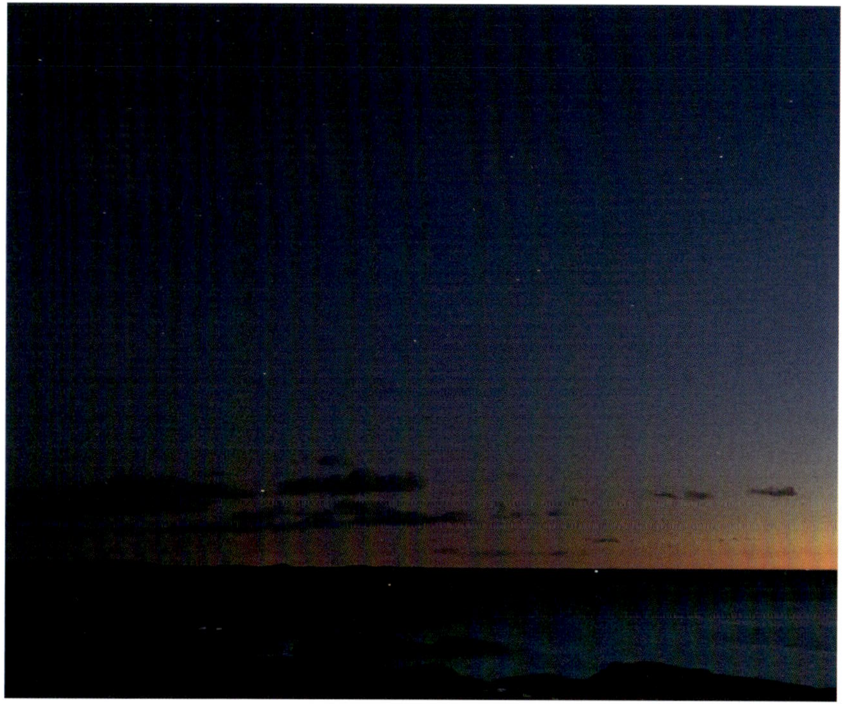

A summer sunset.

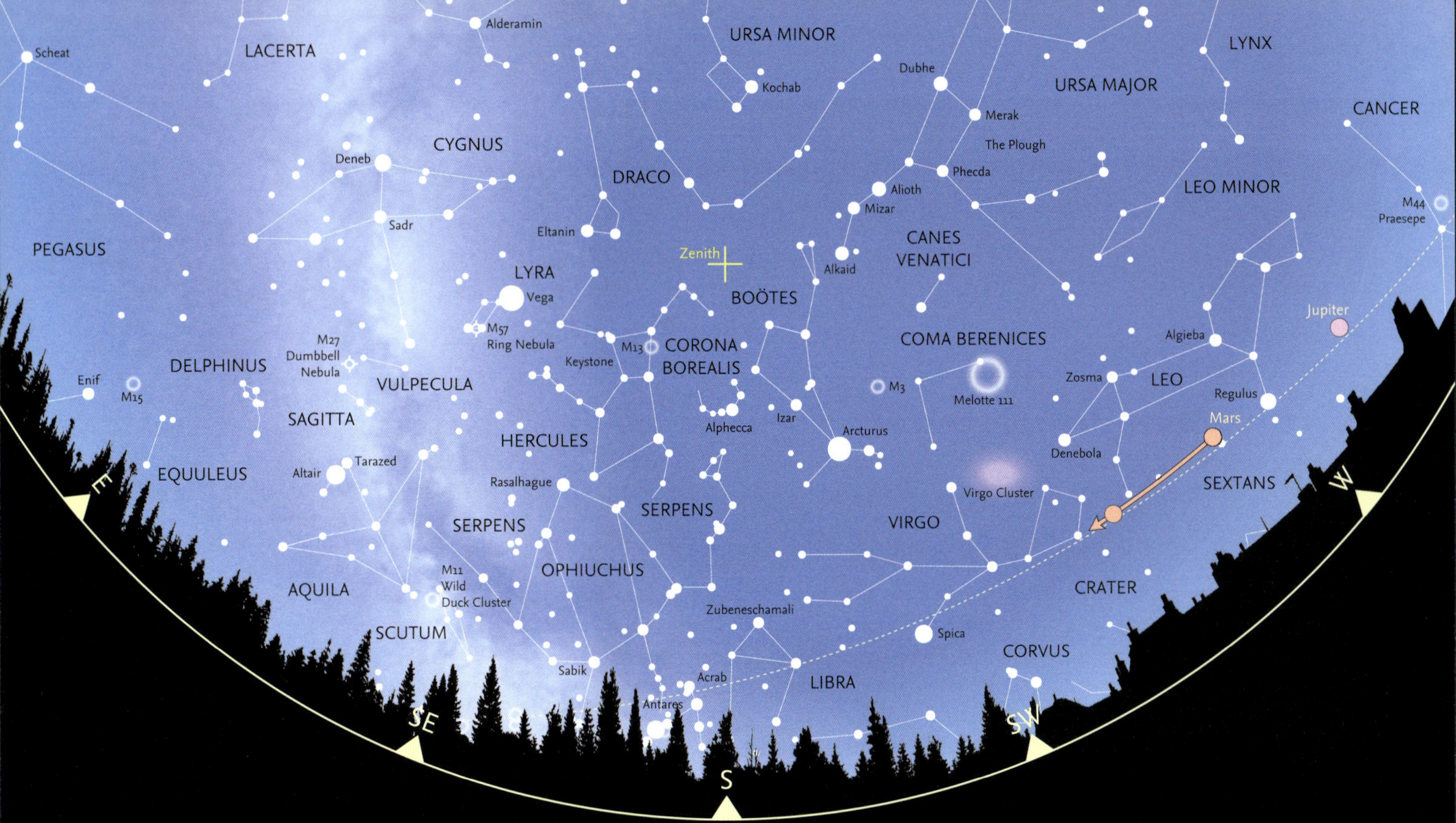

June – Looking South

1 June	00:00 BST (GMT + 1hr)
15 June	23:00 BST (GMT + 1hr)
1 July	22:00 BST (GMT + 1hr)

June – Looking South

Although the persistent twilight makes observing even the southern sky difficult, there is still plenty to see after dark. ***Deneb***, ***Vega*** and ***Altair*** forming the Summer Triangle are the first bright stars to be visible in the east. All three are larger than the Sun, Deneb being 200 times greater in size, and up to 200,000 times more luminous. Lying 100 times further away than the other two stars, we see all three having similar apparent brightness in the summer sky. If Deneb was the same distance as Vega and Altair, it would be so bright in our sky it would cast shadows, like the Full Moon.

The zodiacal constellation of ***Libra*** lies almost due south. The red supergiant star ***Antares*** – the name means the 'rival of Mars' – in ***Scorpius*** is visible slightly to the east of Libra, but the 'tail' or 'sting' remains below the horizon. Higher in the sky is the large constellation of ***Ophiuchus*** (the Serpent Bearer), lying between the two halves of the constellation of ***Serpens***: ***Serpens Caput*** (Head of the Serpent) to the west and ***Serpens Cauda*** (Tail of the Serpent) to the east. The ecliptic runs across Ophiuchus, and the Sun actually spends far more time in the constellation than it does in the 'classical' zodiacal constellation of Scorpius, a small area of which lies between Libra and Ophiuchus.

Higher in the southern sky, the three constellations of ***Boötes***, ***Corona Borealis*** and ***Hercules*** are now better placed for observation than at any other time of the year. This is an ideal time to observe the fine globular cluster of ***M13*** in Hercules with binoculars or a telescope.

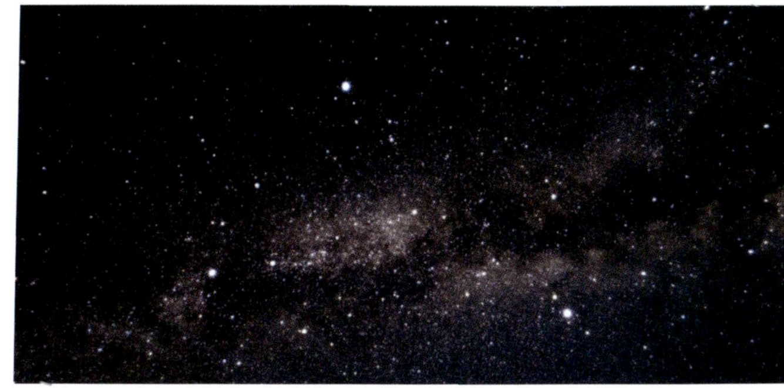

The Summer Triangle: the vertices are marked by the stars Deneb, Vega and Altair.

The Moon's phases for June 2027

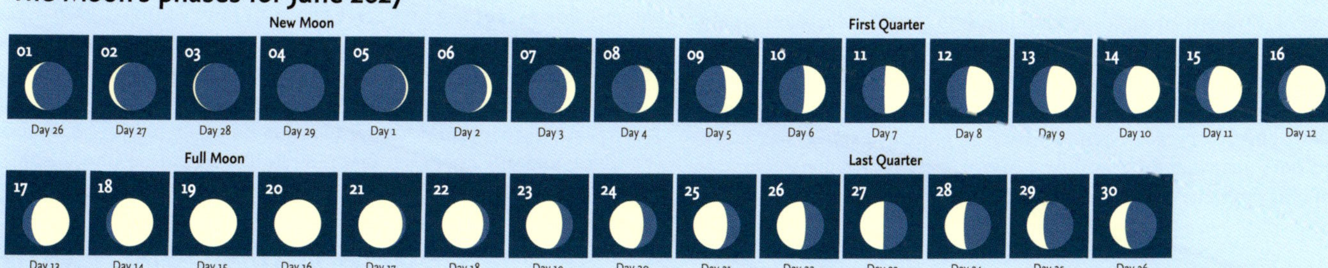

June – Moon and Planets

The Moon

A New Moon occurs on 4 June and two days later ***Mercury*** (mag. 1.4) appears 2.7°S of the waxing crescent Moon, prominent in the sky just before sunset. On 9 June ***Jupiter*** (mag. -1.9) sits 1.9°N of the Moon, the following day it's the turn of ***Mars*** (mag. 0.8) passing 3.9°N of the Moon and ***Regulus*** lies 2.2°N of the Moon. The First Quarter Moon is present on 11 June. On 14 June ***Spica*** lies 3.0°N of the waxing gibbous Moon, three days later ***Antares*** passes less than a tenth of a degree of the Moon, low in the southeast after sunset. The Moon is Full on 19 June and Last Quarter Moon arises on 27 June.

The Planets

Mercury appears just after sunset and switches to the early morning after 23 June. Its magnitude fluctuates, dimming from mag. 0.6 to 5.9 and then brightening to mag. 3.6 at the end of the month. On 6 June Mercury (mag. 1.4) sits 2.7°S of the thin waxing crescent Moon low in the sky. ***Venus*** is visible at dawn at mag. -3.9. On 16 June Venus passes 4.7°N of ***Aldebaran*** in ***Taurus***, a challenging sight through the glow of the sunrise. ***Mars*** stays in ***Leo***, visible after sunset and starting the month at mag. 0.8, dimming to mag. 1.1 at the end of June. On 10 June Mars (mag. 0.8) will appear 3.9°N of the bright waxing crescent Moon. ***Jupiter*** stays in Leo with Mars, visible until midnight at mag. -1.9, dimming slightly to -1.8 as the month progresses. Jupiter (mag. -1.9) will appear 1.9°N of the waxing crescent Moon on 9 June. ***Saturn*** lies in ***Pisces***, rising an hour or so before the Sun at mag. 0.7, brightening slightly to mag. 0.6 towards the end of the month. ***Uranus*** sits in Taurus, setting with the Sun, at mag. 5.8. ***Neptune*** is in Pisces, rising at dawn, at mag. 7.8.

The path of the Sun and the planets along the ecliptic in June.

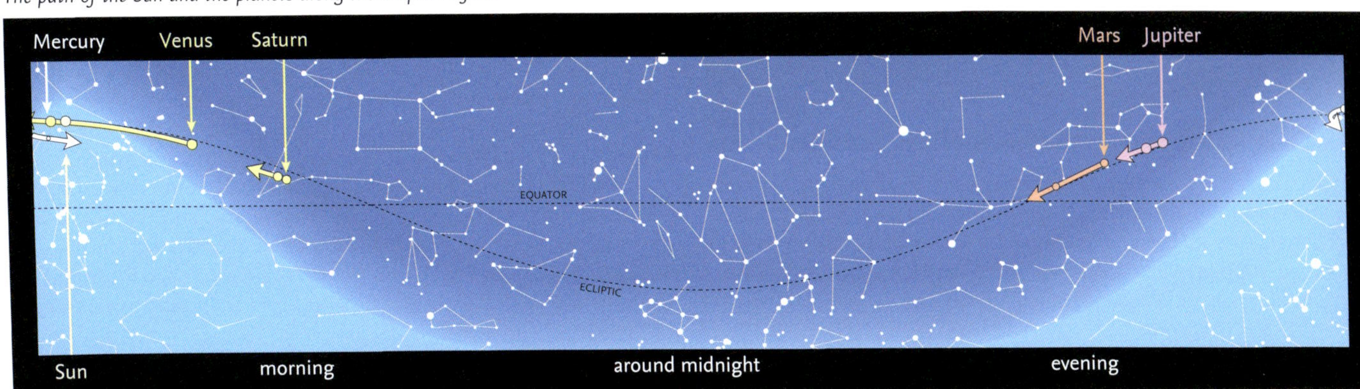

Calendar for June

04	19:40	New Moon
06	06:29	Mercury (mag. 1.4) 2.7°S of the Moon
06	14:54	Moon at perigee = 361,703 km
09	10:54	Jupiter (mag. -1.9) 1.9°N of the Moon
10	01:01	Regulus 2.2°N of the Moon
10	22:45	Mars (mag. 0.8) 3.9°N of the Moon
11	10:56	First Quarter Moon
14	01:01	Spica 3.0°N of the Moon
16	10:11	Venus (mag. -3.9) 4.7°N of Aldebaran
17	13:46	Antares 0.0°S of the Moon
19	00:44	Full Moon
21	14:11	Summer Solstice
22	05:08	Moon at apogee = 405,698 km

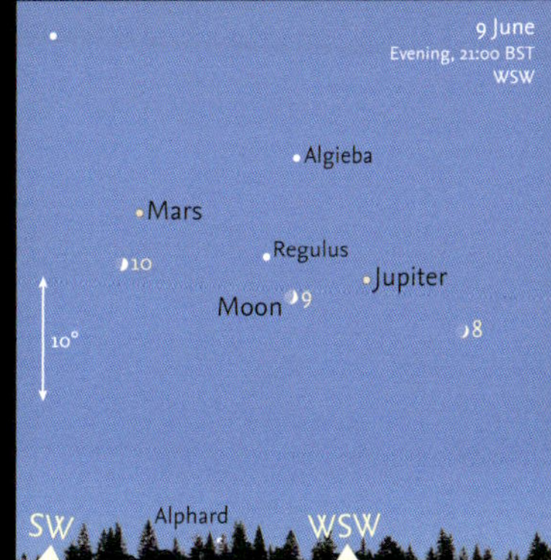

9 June • *The waxing crescent Moon with Jupiter and Mars in Leo.*

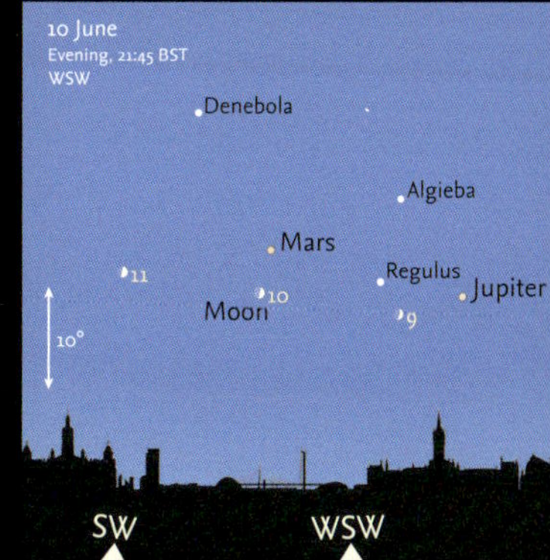

10 June • *The Moon next to Mars just before First Quarter.*

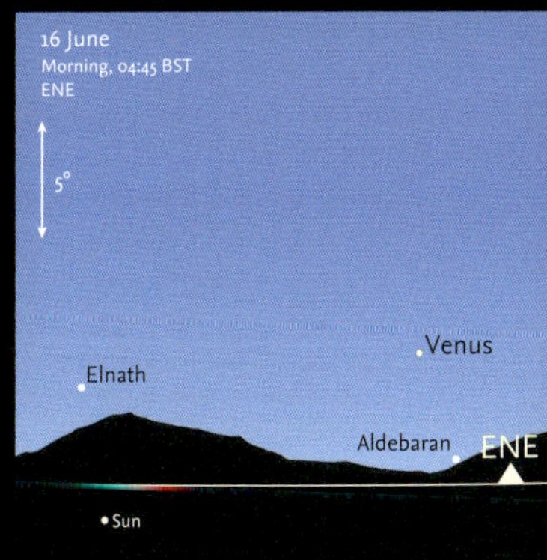

16 June • *Venus in Taurus, close to Aldebaran.*

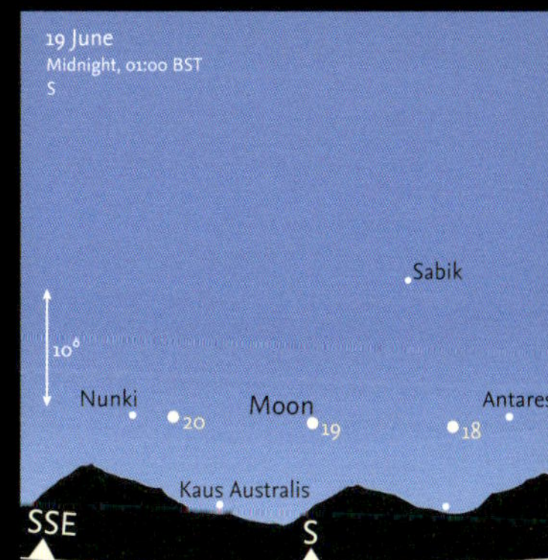

19 June • *Full Moon in Sagittarius.*

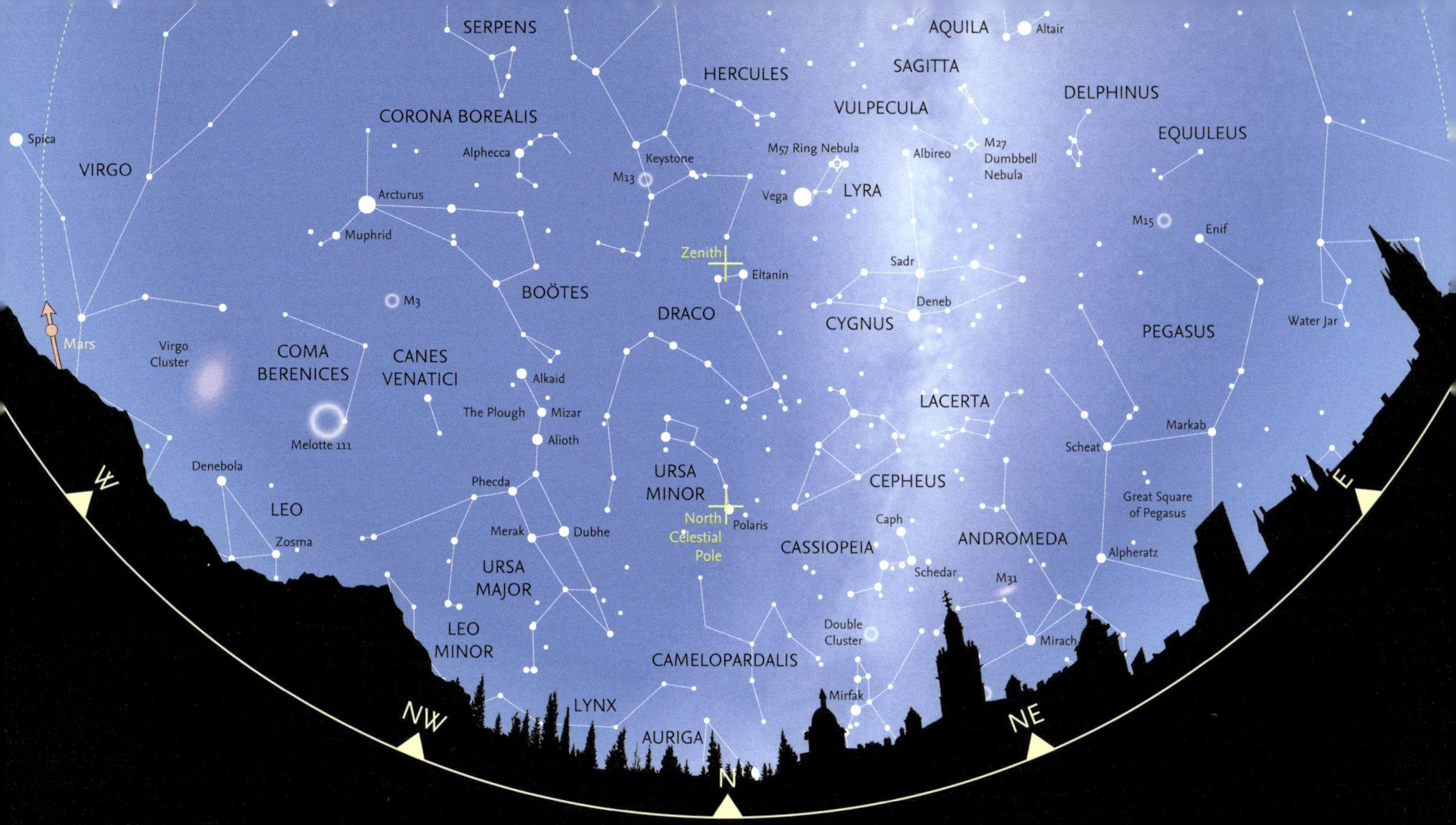

July – Looking North

1 July	00:00 BST (GMT + 1hr)
15 July	23:00 BST (GMT + 1hr)
1 August	22:00 BST (GMT + 1hr)

July – Looking North

As in June, light nights and the chance of observing noctilucent clouds persist throughout July, but later in the month (and particularly after midnight) some of the major constellations are more easily seen. ***Auriga*** is low on the northern horizon. ***Cassiopeia*** is clearly visible in the northeast and ***Perseus***, to its south, is beginning to climb clear of the horizon. The band of the Milky Way, from Perseus through Cassiopeia towards ***Cygnus***, stretches up into the northeastern sky. The star ***Albireo*** (β Cygni) marks the Swan's beak and through binoculars or a telescope it can be seen to be an optical double: the brighter star is an orange giant 100 times more luminous than the Sun; the second star is blue-white and 230 times more luminous than the Sun. The stars are not thought to be orbiting each other (in a binary system).

If the sky is dark and clear, you may be able to make out the small, faint constellation of ***Lacerta***, lying across the Milky Way between Cassiopeia and Cygnus. In the east, the stars of ***Pegasus*** are now well clear of the horizon, with the main line of stars forming ***Andromeda*** roughly parallel to the horizon in the northeast. ***Cepheus*** and ***Ursa Major*** are on opposite sides of ***Polaris*** and ***Ursa Minor***, in the east and west, respectively. The head of ***Draco*** is very close to the zenith so the whole of this winding constellation is readily seen.

Meteors

July brings three meteor showers; there are two minor radiants active in the constellations of ***Capricornus*** and ***Aquarius***. Because of their location, however, observing conditions are not particularly favourable for northern-hemisphere observers. The first shower, the ***α-Capricornids***, active from 3 July–15 August (peaking 30 July), has a maximum rate of about five per hour; however, it does often produce very bright fireballs. The parent body is Comet 169P/NEAT. The most prominent shower is probably that of the ***Southern δ-Aquariids***, which are active from around 12 July–23 August, with a peak on 30 July, although even then the rate is unlikely to reach 25 meteors per hour. In this case, the parent body may be Comet 96P/Machholz. This year, both shower maxima occur when the Moon is a thin waning crescent, meaning there will be very little moonlight present, the Moon rising after midnight. A chart showing the ***δ-Aquariids*** radiant is shown on page 30. From London the radiant rises above the horizon after 21:00 UTC (22:00 local time). The Perseids begin on 17 July and peak on 12–13 August.

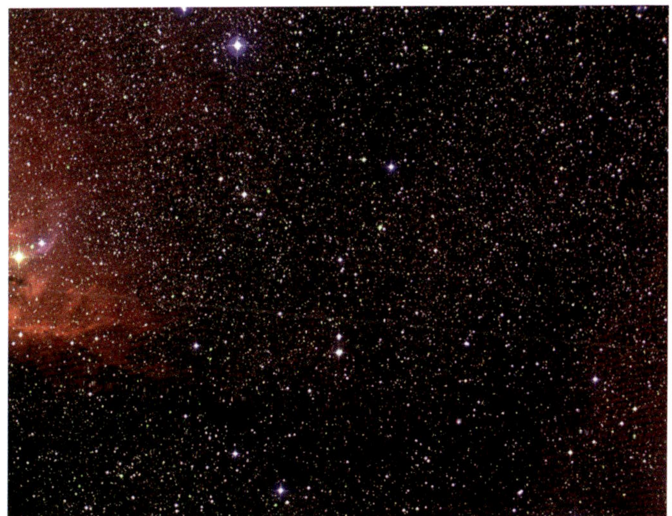

Cygnus the Swan.

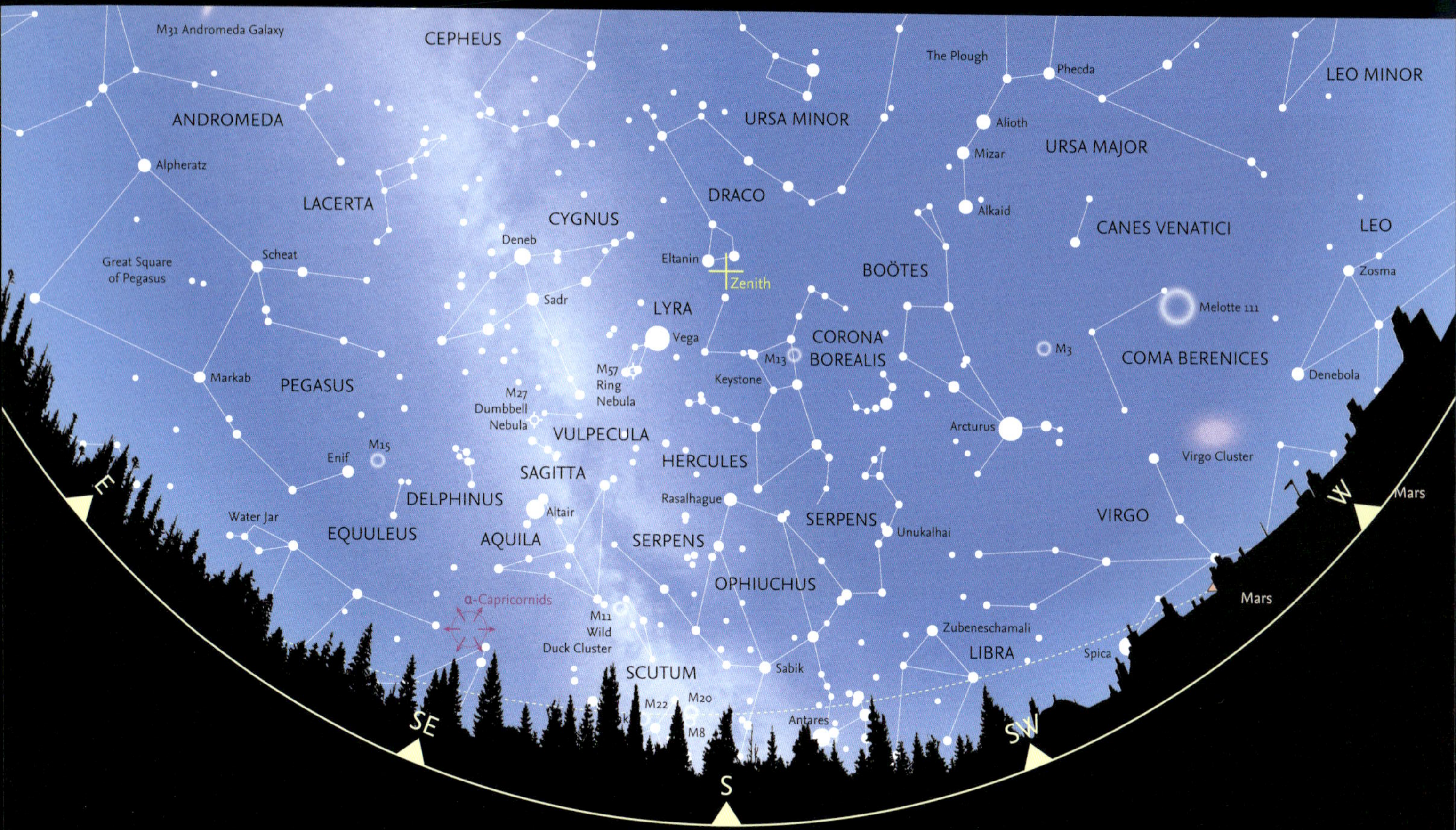

July – Looking South

1 July	00:00 BST (GMT + 1hr)
15 July	23:00 BST (GMT + 1hr)
1 August	22:00 BST (GMT + 1hr)

July – Looking South

Although part of the constellation remains hidden, this is perhaps the best time of year to see ***Scorpius***, with deep red ***Antares*** (α Scorpii), glowing just above the southern horizon. At around midnight (UT, 01:00 BST), part of ***Sagittarius***, with the distinctive asterism of the Teapot, and the dense star clouds of the centre of the Milky Way, are just visible in the south. The Great Dark Rift – actually dust clouds that hide the more distant stars – runs down the Milky Way from ***Cygnus*** towards Sagittarius. Towards its northern end is the small constellation of ***Sagitta*** and the planetary nebula ***M27*** (the Dumbbell Nebula) in the otherwise insignificant constellation of ***Vulpecula***. The sprawling constellation of ***Ophiuchus*** lies close to the meridian for a large part of the month, separating the two halves of ***Serpens***.

In the east, the bright Summer Triangle, consisting of ***Vega*** in ***Lyra***, ***Deneb*** in Cygnus and ***Altair*** in ***Aquila***, begins to dominate the southern sky, as it will throughout August and into September. The small constellation of Lyra, containing Vega (the fifth brightest star in the night sky) and a distinctive quadrilateral of stars to its east and south, lies not far south of the zenith.

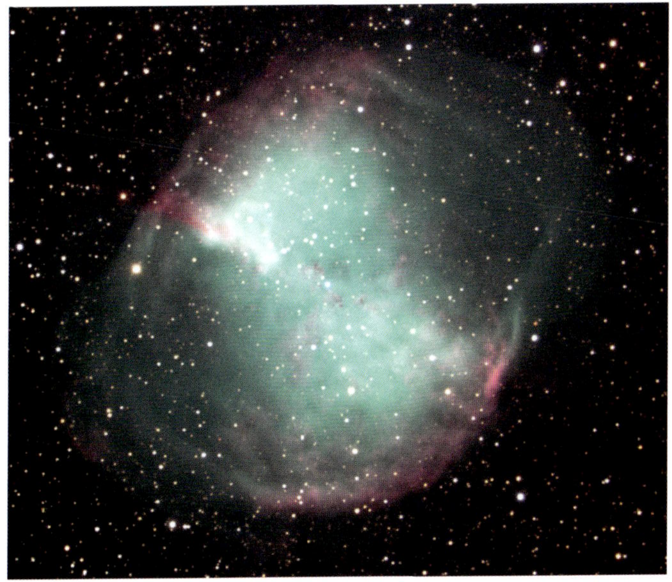

The Dumbbell Nebula, a planetary nebula – the remnant of a dying star.

The Moon's phases for July 2027

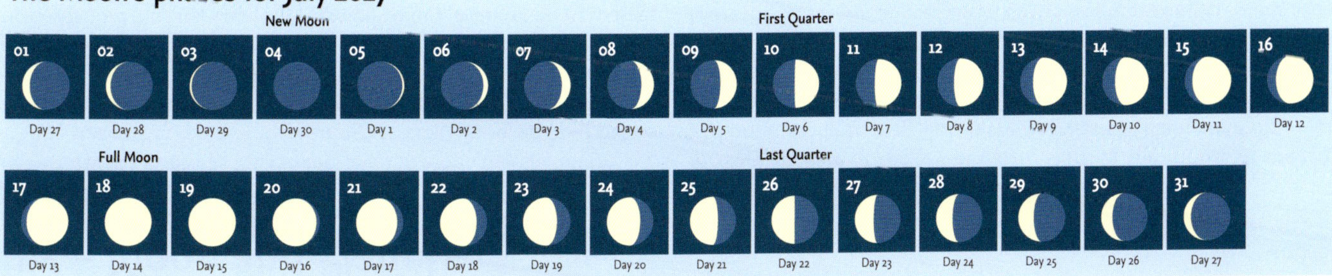

July – Moon and Planets

The Moon

The first day of July brings the ***Pleiades*** and the Moon together in the sky, with the star cluster 0.9°S of the waning crescent Moon, an hour before sunrise. The Moon becomes New on 4 July, on 7 July the waxing crescent Moon sits 2.5°S of the gas giant ***Jupiter*** (mag. -1.8), with ***Regulus*** in ***Leo*** lying 2.3°N, all three dip below the horizon a few hours before midnight. The day before the First Quarter Moon on 10 July, ***Mars*** (mag. 1.1) passes 4.7°N of the Moon and on 11 July ***Spica*** is 3.2°N of the waxing gibbous Moon. On 14 July the 'rival of Mars', ***Antares*** is positioned 0.1°N of the waxing gibbous Moon. The Moon is Full on 18 July and on 26 July the Moon reaches Last Quarter. On 28 July the beautiful ***Pleiades*** are back close to the waning crescent Moon, situated only 1.0°S.

The Planets

Mercury appears in the sky at dawn, starting the month at mag. 3.3 and brightening to mag. -1.3. It reaches western elongation on 16 July. ***Venus*** continues to rise in the early hours at mag. -3.9. ***Mars*** continues in ***Leo*** until 4 July, after which it progresses into ***Virgo***, visible after sunset. It starts the month at mag. 1.1, gradually dimming to mag. 1.2. Mars (mag. 1.1) will appear 4.7°N of the bright waxing crescent Moon on 9 July. ***Jupiter*** stays in Leo with Mars, setting a few hours before midnight at mag. -1.8, dimming to -1.7 as the month progresses. On 7 July Jupiter will be 2.5°N of the waxing crescent Moon. It lies 0.4°N of ***Regulus*** on 25 July at mag. -1.7. ***Saturn*** lies in ***Pisces***, appearing a few hours before sunrise, slowly brightening to mag. 0.5. ***Uranus*** sits in ***Taurus***, climbing above the horizon an hour before sunrise, at mag. 5.8. ***Neptune*** is in Pisces rising shortly after midnight, brightening from mag. 7.8 to 7.7. It enters retrograde motion on 10 July.

The path of the Sun and the planets along the ecliptic in July.

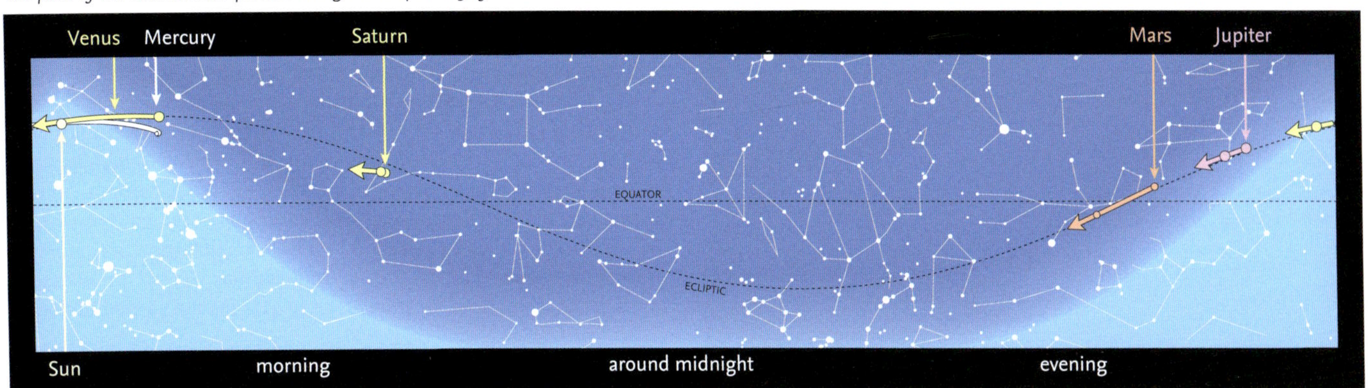

Calendar for July

01	07:04	Pleiades 0.9°S of the Moon
04	03:02	New Moon
04	20:55	Moon at perigee = 358,284 km
05	05:05	Earth at aphelion (152,100,643 km = 1.01673 AU)
07	03:14	Jupiter (mag. -1.8) 2.5°N of the Moon
07	09:01	Regulus 2.3°N of the Moon
09	07:52	Mars (mag. 1.1) 4.7°N of the Moon
10	00:30	Neptune enters retrograde motion
10	18:39	First Quarter Moon
11	06:50	Spica 3.2°N of the Moon
14	19:34	Antares 0.1°N of the Moon
16	03:31	Mercury at greatest elongation (20.7°W, mag. 0.4)
18	15:45	Full Moon
19	11:52	Moon at apogee = 406,217 km
25	06:23	Jupiter (mag. -1.7) 0.4°N of Regulus
26	16:55	Last Quarter Moon
28	16:20	Pleiades 1.0°S of the Moon
30		Delta (δ) Aquariid meteor shower maximum
30		Alpha (α) Capricornid meteor shower maximum

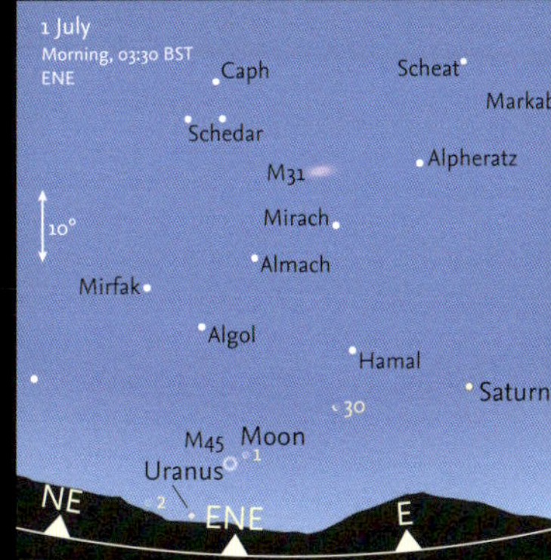

1 July • *The waning crescent Moon lies next to the Pleiades (M45). Uranus and Saturn are nearby.*

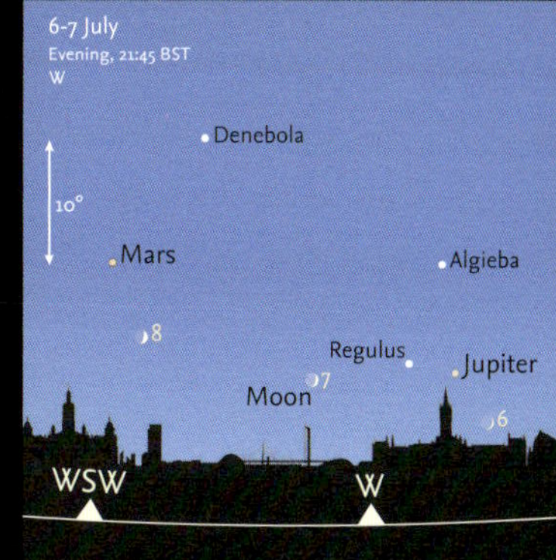

6–7 July • *Waxing crescent Moon passes by Jupiter.*

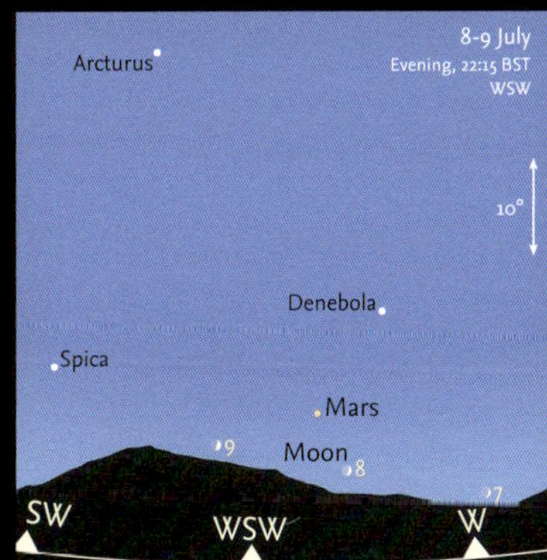

8–9 July • *The Moon swings past Mars.*

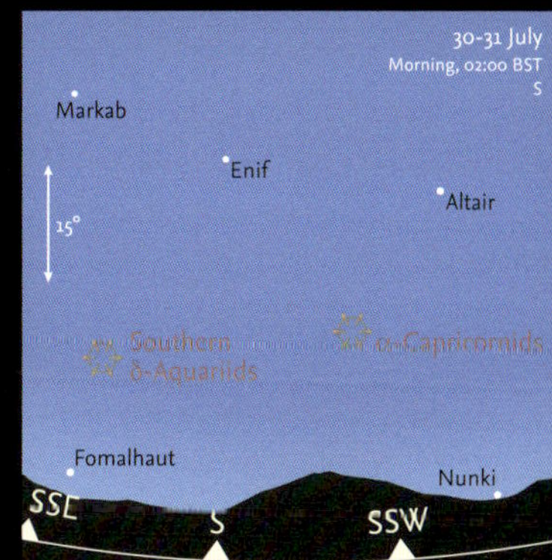

30–31 July • *The peak of the Southern δ-Aquariids.*

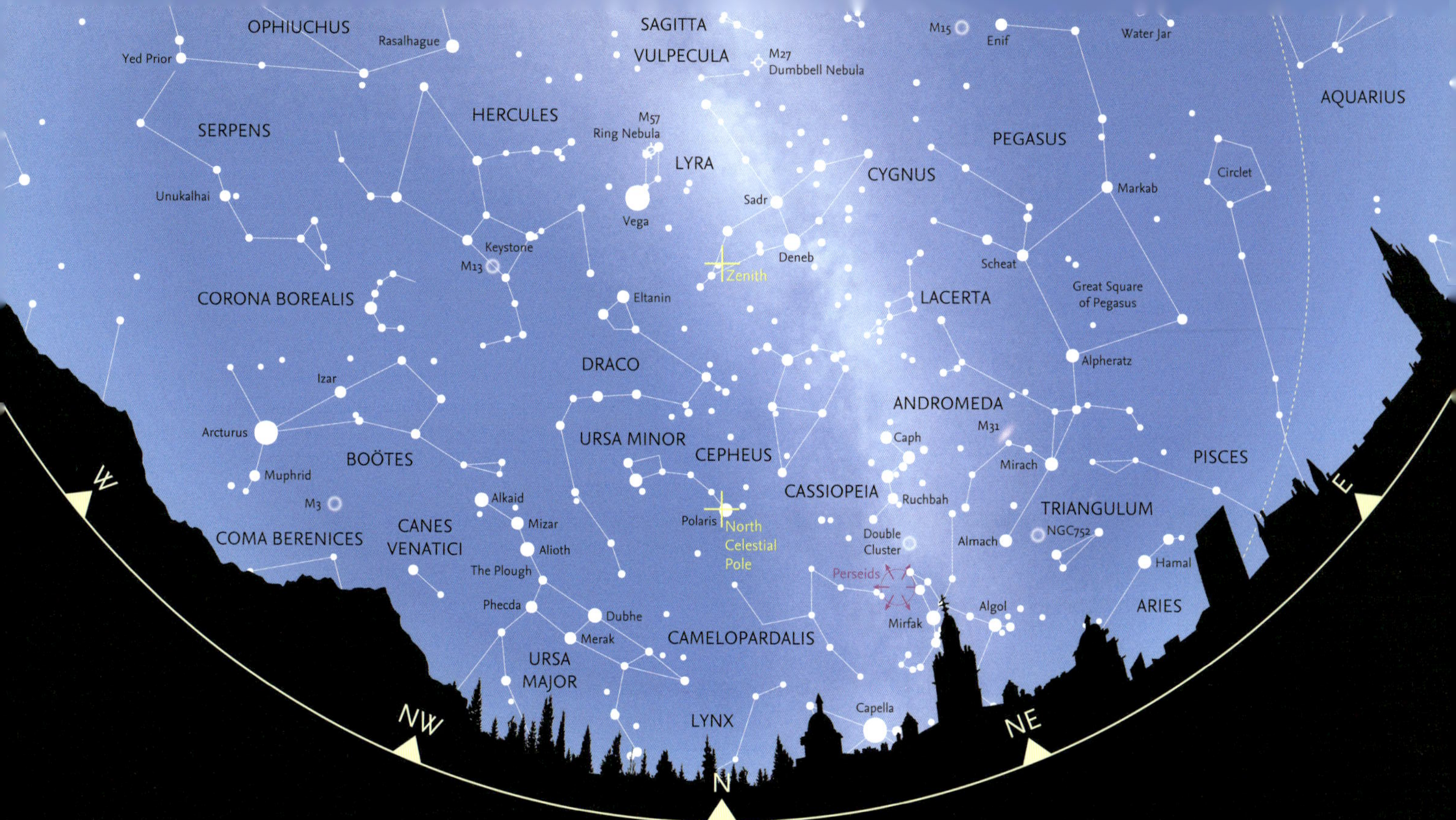

August – Looking North

1 August	00:00 BST (GMT + 1hr)
15 August	23:00 BST (GMT + 1hr)
1 September	22:00 BST (GMT + 1hr)

August – Looking North

Ursa Major is now the 'right way up' in the northwest, although some of the fainter stars in the south of the constellation are difficult to see. Beyond it, ***Boötes*** stands almost vertically in the west, but pale orange ***Arcturus*** is sinking towards the horizon. Higher in the sky, both ***Corona Borealis*** and ***Hercules*** are clearly visible, along with ***Draco***.

In the northeast, ***Capella*** is clearly seen, but most of ***Auriga*** still remains below the horizon. Higher in the sky, ***Perseus*** is gradually coming into full view and, later in the night and later in the month, the beautiful ***Pleiades*** cluster rises above the northeastern horizon. Between Perseus and ***Polaris*** lies the faint constellation of ***Camelopardalis***.

Higher still, both ***Cassiopeia*** and ***Cepheus*** are well placed for observation, despite the fact that Cassiopeia is completely immersed in the band of the Milky Way, as is the 'base' of Cepheus. ***Pegasus*** and ***Andromeda*** are now well above the eastern horizon and, below them, the constellation of ***Pisces*** is climbing into view. Two of the stars in the Summer Triangle, ***Deneb*** and ***Vega***, are close to the zenith high overhead.

Meteors

August is the month when one of the best meteor showers of the year occurs: the ***Perseids***. This is a long shower, generally beginning about 17 July and continuing until around 24 August. The peak of the shower occurs over the night of 12–13 August, when the rate may reach as high as 100 meteors per hour (and on rare occasions, even higher). In 2027, there is a bright waxing gibbous at the maximum so the moonlight will interfere with observations before midnight – meteors are generally best seen after this time and skies will be darker for this shower. The Perseids are debris from Comet 109P/Swift-Tuttle (the Great Comet of 1862). Perseid meteors are fast and many of the brighter ones leave persistent trains. Some bright fireballs also occur during the shower, arising from larger chunks of cometary debris.

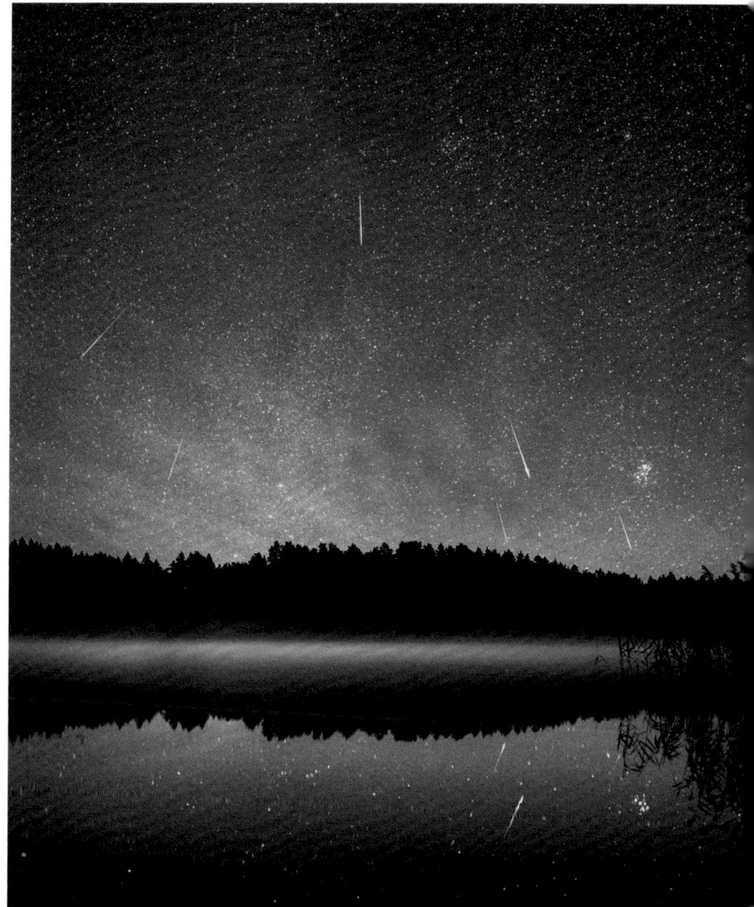

The summer Perseid meteor shower, known for its bright fireballs.

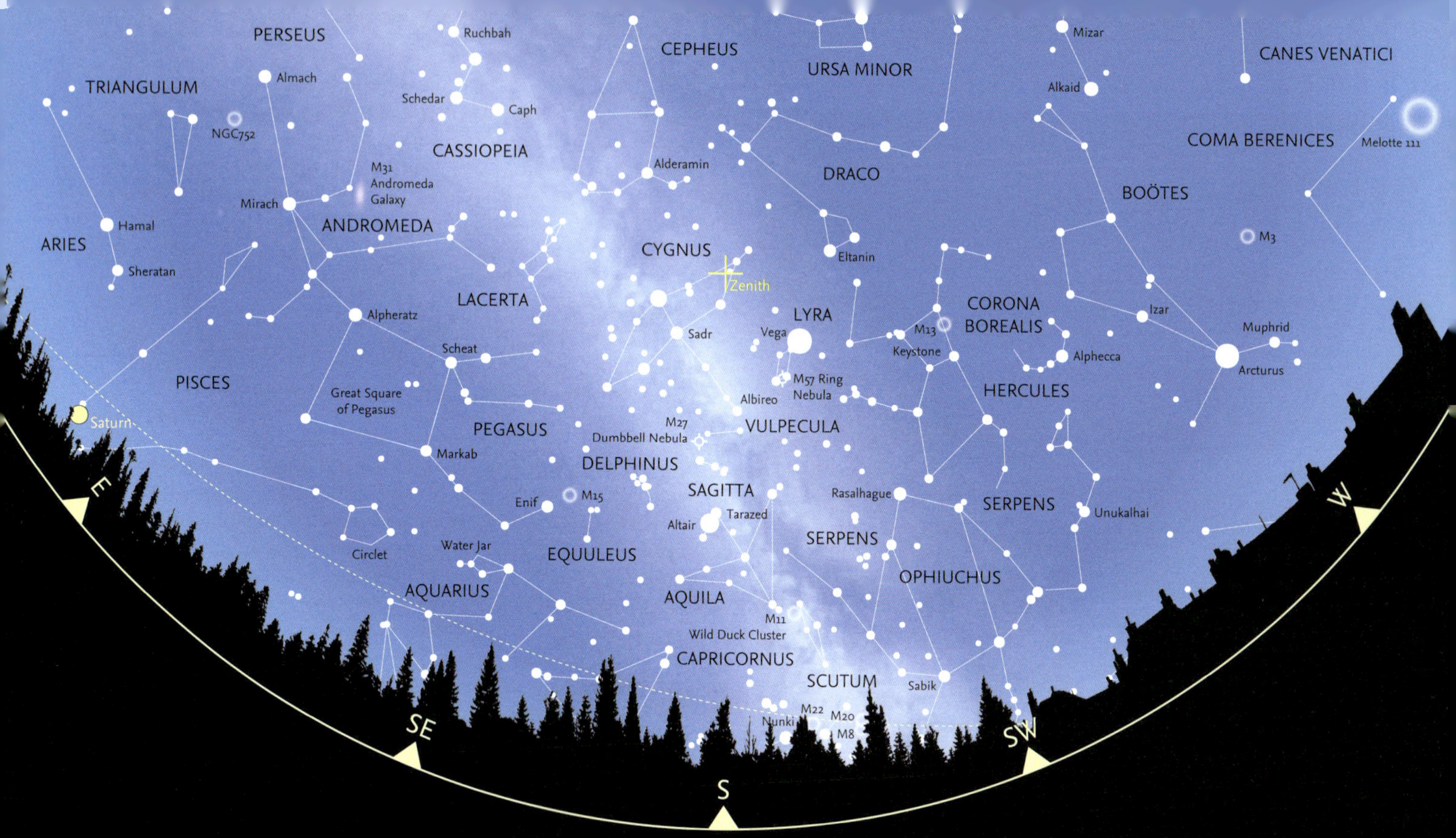

August – Looking South

1 August	00:00 BST (GMT + 1hr)
15 August	23:00 BST (GMT + 1hr)
1 September	22:00 BST (GMT + 1hr)

August – Looking South

The whole of the summer Milky Way stretches across the sky in the south, from ***Cygnus***, high in the sky near the zenith, past ***Aquila***, with bright ***Altair*** (α Aquilae), to part of the constellation of ***Sagittarius*** close to the horizon, where the pattern of stars known as the Teapot is visible. This area contains many nebulae and both open and globular clusters. The star-forming Lagoon Nebula, ***M8*** is close to γ Sagittarii and it is visible to the naked eye on a clear dark night. ***M20***, the Trifid Nebula is close to M8 and can be seen with a small telescope.

Between ***Albireo*** (β Cygni) and Altair lie the two small constellations of ***Vulpecula*** and ***Sagitta***, with the latter easier to distinguish (because of its shape) from the clouds of the Milky Way. One third of the way between Altair and ***Vega*** lies the Coathanger asterism; binoculars are required to see these stars. Between Sagitta and ***Pegasus*** to the east lie the highly distinctive five stars that form the tiny constellation of ***Delphinus*** (the Dolphin), again, one of the few constellations that actually bears some resemblance to the creature after which it is named. Below Aquila, mainly in the star clouds of the Milky Way, lies ***Scutum***, most famous for the bright open cluster, ***M11*** or the Wild Duck Cluster, readily visible in binoculars. To the southeast of Aquila lie the two zodiacal constellations of ***Capricornus*** and ***Aquarius***.

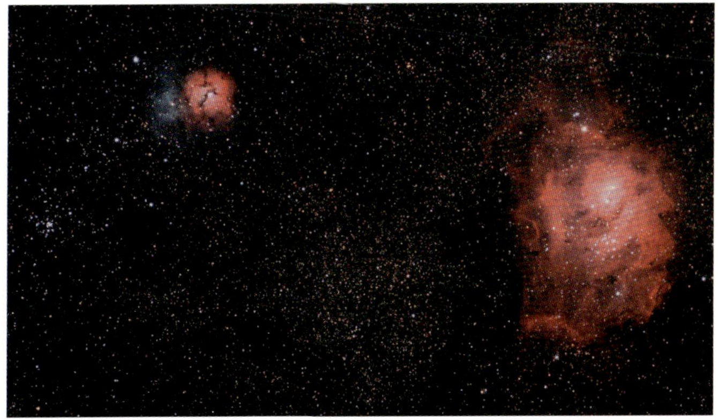

The Lagoon Nebula, a hot, glowing star-forming gas cloud.

The Moon's phases for August 2027

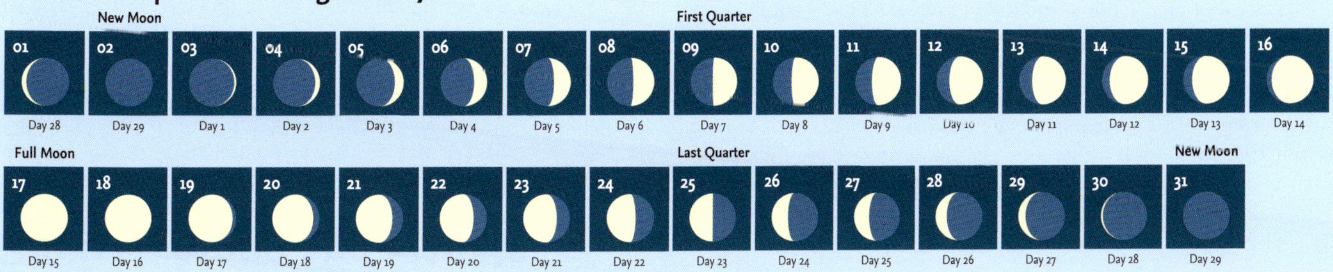

August – Moon and Planets

The Moon

The New Moon arises on 2 August, the thin crescent passes 2.4°S of ***Regulus*** and 3.0°S of dazzling ***Jupiter*** (mag. -1.7) on 3 August. On 7 August ***Spica*** can be seen 3.3°N of the waxing crescent Moon. Two days later the Moon is at First Quarter. On 11 August beautiful red-orange ***Antares*** lies only 0.1°N of the waxing gibbous Moon, visible just after sunset. The Full Moon outshines the sky on 17 August. The waning gibbous Moon passes 0.9°N of the ***Pleiades*** on 24 August, the following day it reaches Last Quarter and the month ends with a New Moon on 31 August.

The Planets

Mercury rises with the Sun until 11 August after which it swings into the evening sky, appearing just after sunset. It brightens from -1.3 to -2.0 and fades to -0.4 at the end of August. ***Venus*** is visible at dawn, rising with the Sun on 12 August after which it trails the Sun and can be seen in the evening sky, still spectacular at mag. -3.9. ***Mars*** is placed in ***Virgo***, visible in the early evening at mag. 1.3 all month. Mars passes 1.7°N of ***Spica*** on 23 August. ***Jupiter*** continues in ***Leo***, setting with the Sun towards the end of the month, at mag. -1.7. Jupiter will appear 3.0°N of the waxing crescent Moon on 3 August, with ***Regulus*** close by. ***Saturn*** lies in ***Pisces***, visible after midnight, starting the month at mag. 0.5 and slowly brightening to mag. 0.3. It enters retrograde motion on 9 August. ***Uranus*** stays in ***Taurus***, appearing shortly after midnight, starting at mag. 5.8 and brightening gradually to 5.7. ***Neptune*** is in Pisces rising an hour or so before midnight, at mag. 7.7.

The path of the Sun and the planets along the ecliptic in August.

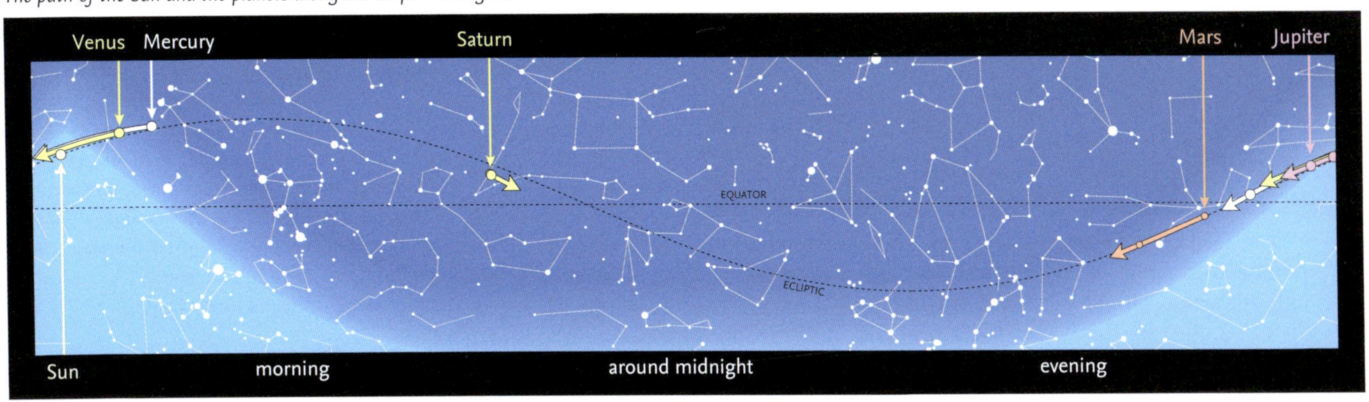

Calendar for August

02	06:28	Moon at perigee = 357,362 km
02	10:05	New Moon
02	10:06	Total solar eclipse (northern Africa, southern Saudi Arabia, Egypt). Partial eclipse visible from UK, Portugal, Spain, Madeira, Gibraltar, Sudan
03	19:00	Regulus 2.4°N of the Moon
03	22:38	Jupiter (mag. -1.7) 3.0°N of the Moon
07	14:03	Spica 3.3°N of the Moon
09	04:54	First Quarter Moon
09	18:55	Saturn enters retrograde motion
11	01:31	Antares 0.1°N of the Moon
12–13		Perseid meteor shower maximum
15	14:23	Moon at apogee = 406,086 km
17	07:13	Penumbral lunar eclipse. Visible from the Americas, Alaska and Antarctica
17	07:29	Full Moon
23	14:36	Mars (mag. 1.3) 1.7°N of Spica
24	23:36	Pleiades 0.9°S of the Moon
25	02:27	Last Quarter Moon
30	15:36	Moon at perigee = 359,208 km
31	17:41	New Moon

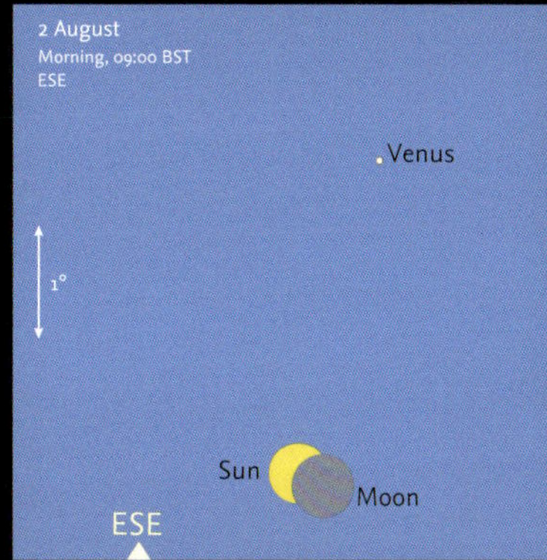

2 August • *Partial solar eclipse visible from London at 08:00 UT (09:00 BST).*

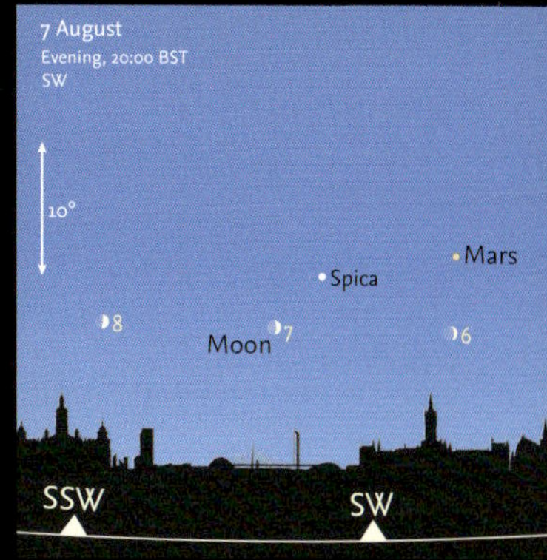

7 August • *Spica and the waxing crescent Moon, Mars nearby.*

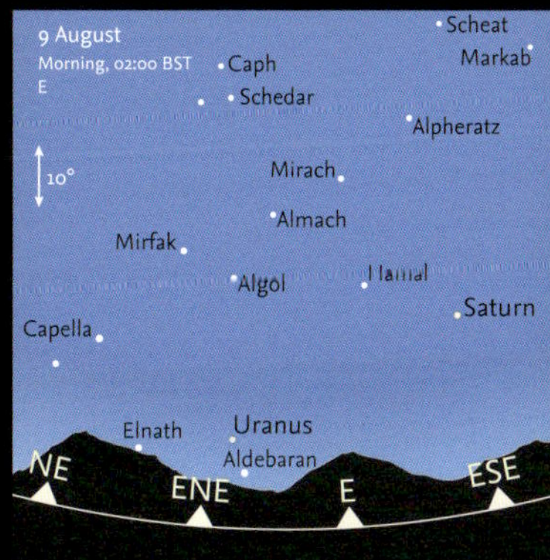

9 August • *Saturn enters retrograde motion. Uranus is close by.*

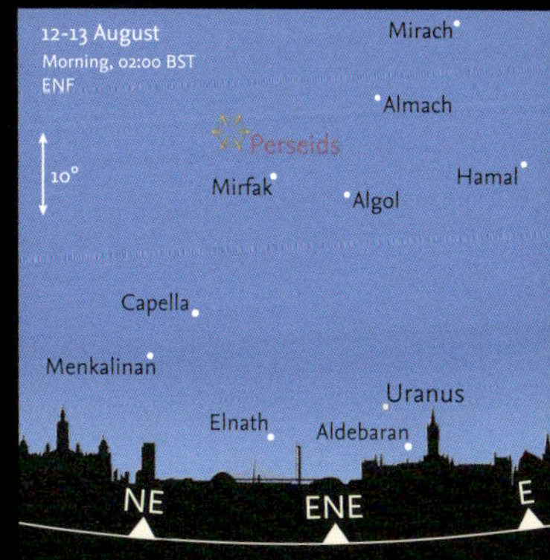

12–13 August • *The peak of the Perseid meteor shower.*

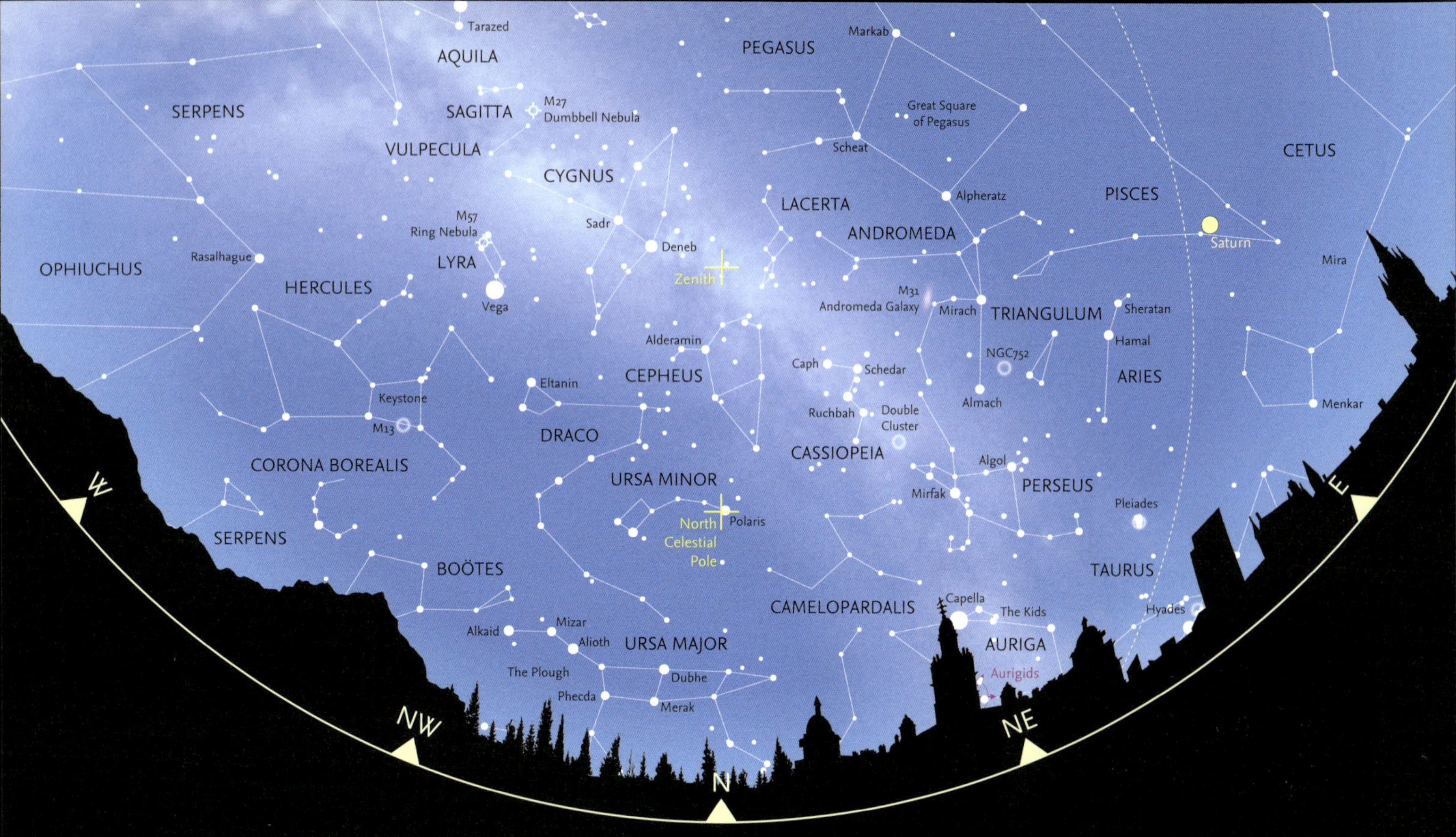

September – Looking North

1 September	00:00 BST (GMT + 1hr)
15 September	23:00 BST (GMT + 1hr)
1 October	22:00 BST (GMT + 1hr)

September – Looking North

The (northern) autumnal equinox occurs on 23 September, when the Sun moves south of the equator in Virgo.

Ursa Major is now low in the north and to the northwest ***Arcturus*** and much of ***Boötes*** sink below the horizon later in the night and later in the month. In the northeast, ***Auriga*** is beginning to climb higher in the sky. Later in the month, ***Taurus***, with orange ***Aldebaran*** (α Tauri), and even ***Gemini***, with ***Castor*** and ***Pollux***, become visible in the east and northeast just after midnight. Due east, ***Andromeda*** is now clearly visible, with the small constellations of ***Triangulum*** and ***Aries*** (the latter a zodiacal constellation) directly below it. Practically the whole of the northern Milky Way is visible, arching across the sky, both in the north and in the south. The clouds of stars are easier to see in ***Cassiopeia*** and ***Cygnus***. The ***Double Cluster*** in ***Perseus*** is well placed for observation. ***Cepheus*** is 'upside down' near the zenith, and the head of ***Draco*** and ***Hercules*** (with the globular cluster ***M13***) beyond it are clearly visible.

Meteors

There are two minor showers in September. The ***α-Aurigids*** starts 28 August and ends 5 September and tends to have two peaks of activity. The principal peak occurs on 1 September, a day after the New Moon so viewing conditions will be favourable. At maximum, however, the rate only reaches around five meteors per hour, although the meteors are bright and relatively easy to photograph. The ***Southern Taurid*** shower begins this month (on 10 September) and, although rates are low, often produces very bright fireballs. This is a very long shower, lasting until about 20 November. As a slight compensation for the lack of shower activity, however, in September the number of sporadic meteors reaches its highest rate of the year.

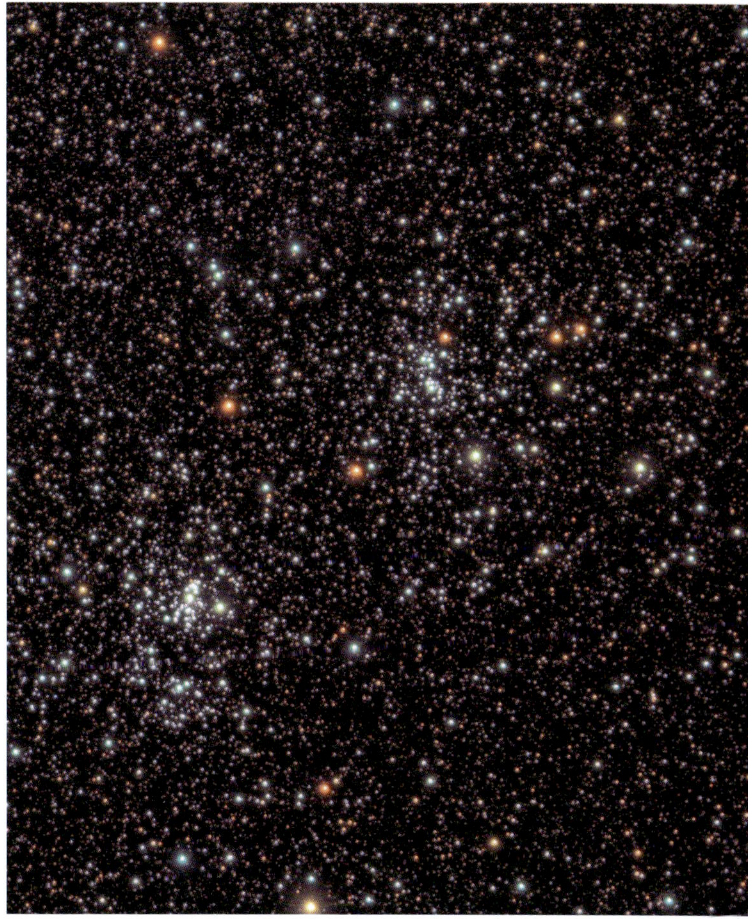

The Double Cluster, composed of two star clusters 7,500 light years away.

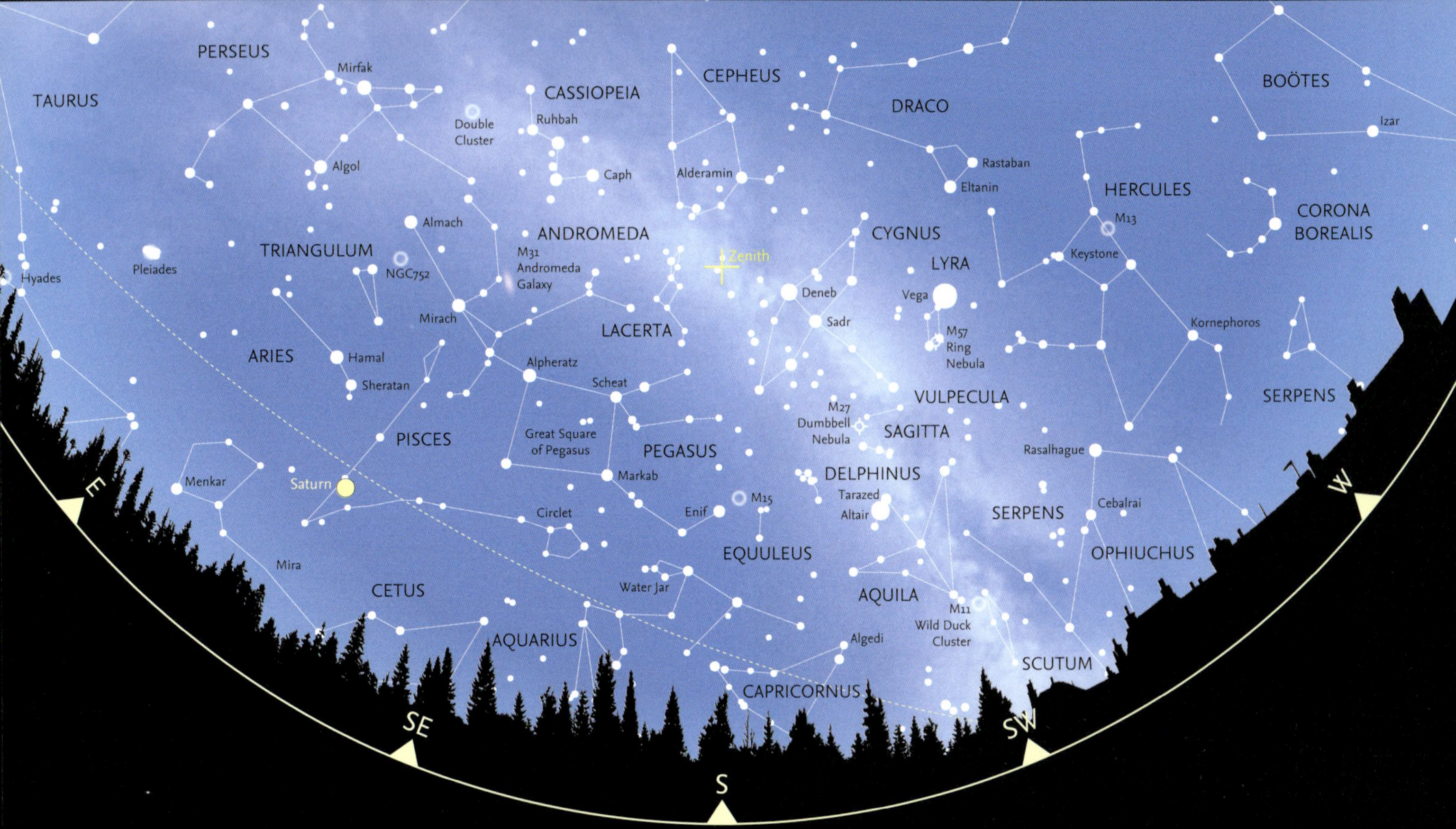

September – Looking South

1 September	00:00 BST (GMT + 1hr)
15 September	23:00 BST (GMT + 1hr)
1 October	22:00 BST (GMT + 1hr)

September – Looking South

The Summer Triangle is now high in the southwest, with the Great Square of ***Pegasus*** high in the southeast. Below Pegasus are the two zodiacal constellations of ***Capricornus*** and ***Aquarius***. ***Algedi*** (α Capricorni) is actually an optical double, with the two stars (α^1 Cap and α^2 Cap) readily seen with the naked eye. ***Dabih*** (β Capricorni), just to the south, is also a double star and the components are relatively easy to separate with binoculars. In Aquarius, there is a small asterism consisting of four stars, resembling a tiny letter 'Y', known as the Water Jar. Below Aquarius is a sparsely populated area of the sky with just one bright star in the constellation of ***Piscis Austrinus***, barely visible above the horizon. In classical illustrations, water is shown flowing from the Water Jar towards bright ***Fomalhaut*** (α Piscis Austrini).

Another zodiacal constellation, ***Pisces***, is now clearly visible to the east of Aquarius. Although faint, there is a distinctive asterism of stars, known as the Circlet, south of the Great Square and another line of faint stars to the east of Pegasus. Still farther down towards the horizon is the constellation of ***Cetus***, with the famous variable red giant star ***Mira*** (o Ceti) at its centre. When Mira is at maximum brightness (around mag. 3.5) it is clearly visible to the naked eye, but it disappears as it fades towards minimum (about mag. 9.5 or less).

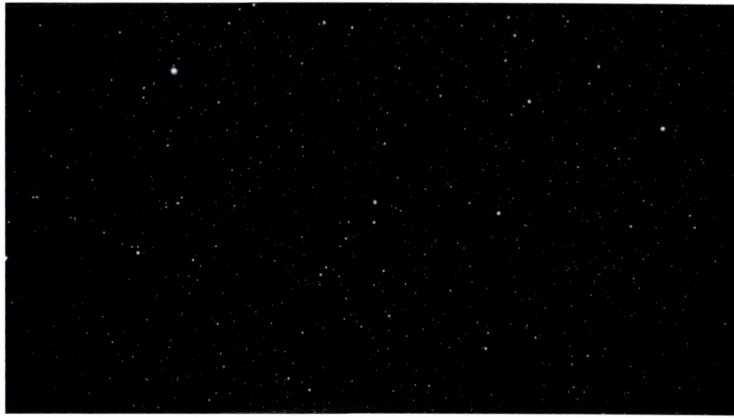

The constellation Piscis Austrinus and its brightest star, Fomalhaut.

The Moon's phases for September 2027

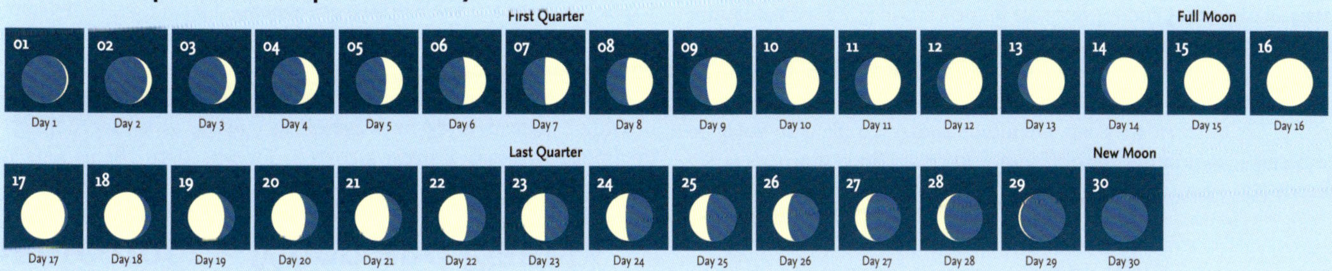

September – Moon and Planets

The Moon

Mercury (mag. -0.3) sits 4.5°N of the waxing crescent Moon on 2 September, with ***Virgo***'s brightest star ***Spica*** placed 3.2°N on 3 September, all above the horizon just after sunset. ***Antares*** lies 0.0°N of the First Quarter Moon on 7 September. The Moon becomes Full on 15 September. In the early hours of 21 September the waning gibbous Moon passes 0.8°N of the ***Pleiades***, two days later it arrives at Last Quarter. On 27 September ***Regulus*** can be found 2.4°N of the waning crescent Moon, the following day ***Jupiter*** (mag. -1.7) sits 4.2°N of the Moon just before sunrise. The end of the month is marked by a New Moon.

The Planets

Mercury can be seen in the early evening just after sunset. It reaches eastern elongation on 24 September. It dims over the course of the month from -0.4 to 0.2. On 2 September Mercury (mag. -0.3) appears 4.5°N of the waxing crescent Moon. On 21 September it lies 0.2°S of ***Spica***, at mag. 0.1. ***Venus*** dominates the early evening sky at mag. -3.9. ***Mars*** progresses from ***Virgo*** to ***Libra*** after 15 September, visible in the early evening at mag. 1.3, slowly fading to 1.4. ***Jupiter*** stays in ***Leo***, setting with the Sun, at mag. -1.7 all month. On 28 September Jupiter passes 4.2°N of the waning crescent Moon. ***Saturn*** lies in ***Pisces***, climbing above the horizon late in the evening. It starts the month at mag. 0.3 and slowly brightens to mag. 0.2. ***Uranus*** stays in ***Taurus***, visible from midnight at mag. 5.7 and brightening gradually to 5.6. The ice giant enters retrograde motion on 15 September. ***Neptune*** is in Pisces visible from late evening, at mag. 7.7. Our farthest planet reaches opposition on 28 September.

The path of the Sun and the planets along the ecliptic in September.

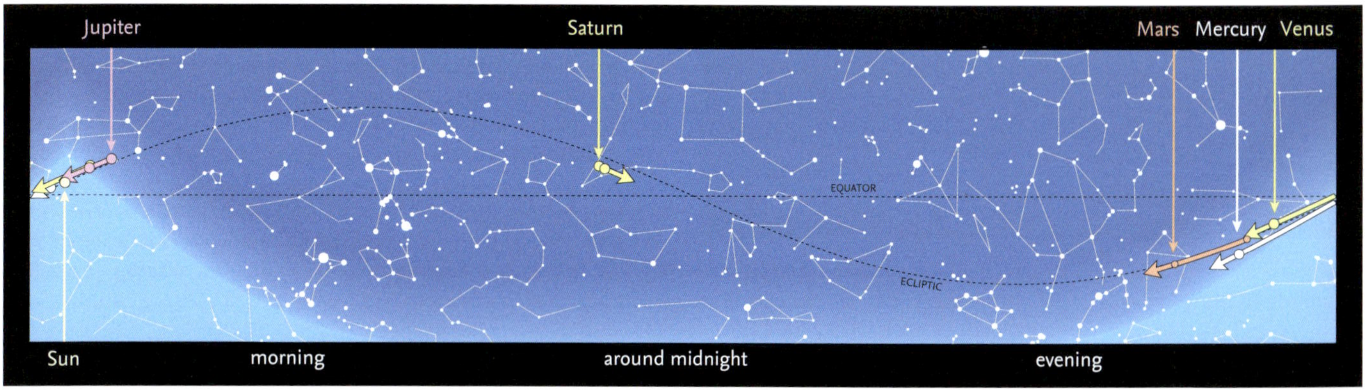

Calendar for September

02	04:46	Mercury (mag. -0.3) 4.5°N of the Moon
03	23:07	Spica 3.2°N of the Moon
07	08:34	Antares 0.0°N of the Moon
07	18:31	First Quarter Moon
11	23:39	Moon at apogee = 405,384 km
15	11:16	Uranus enters retrograde motion
15	23:04	Full Moon
21	03:21	Mercury (mag. 0.1) 0.2°S of Spica
21	05:12	Pleiades 0.8°S of the Moon
23	06:02	Autumnal Equinox
23	10:20	Last Quarter Moon
24	16:59	Mercury at greatest elongation (26.3°E, mag. 0.1)
27	15:35	Regulus 2.4°N of the Moon
27	20:12	Moon at perigee = 363,455 km
28	15:10	Neptune (mag. 7.7) at opposition
28	15:39	Jupiter (mag. -1.7) 4.2°N of the Moon

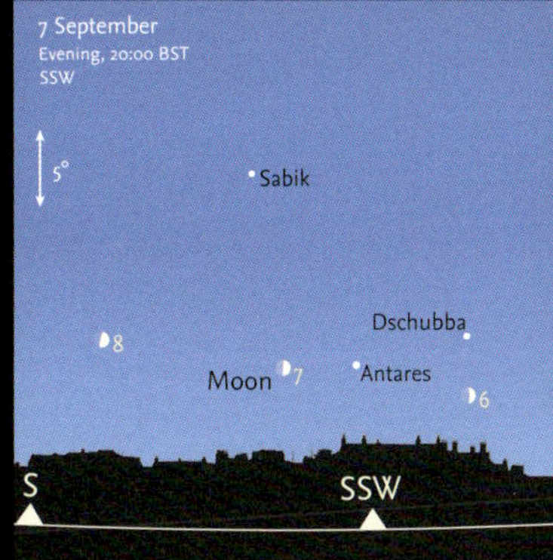

7 September • *Antares and the First Quarter Moon together in the southwest.*

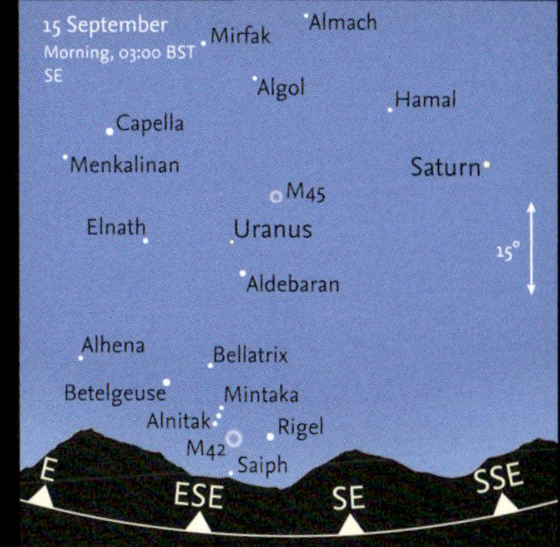

15 September • *Uranus enters retrograde motion in Taurus, the Pleiades (M45) and Aldebaran nearby.*

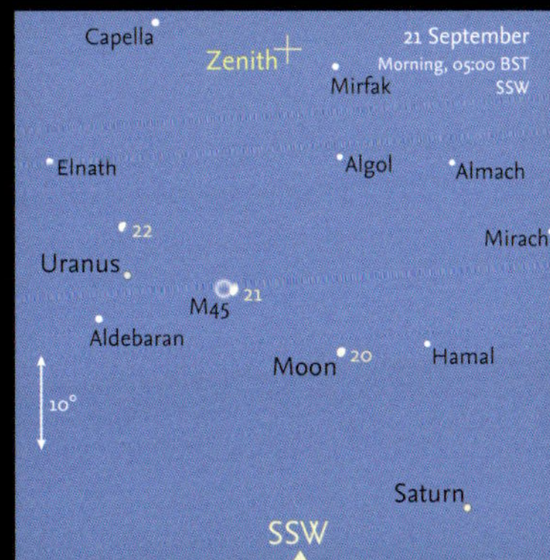

21 September • *The Pleiades and the waning gibbous Moon.*

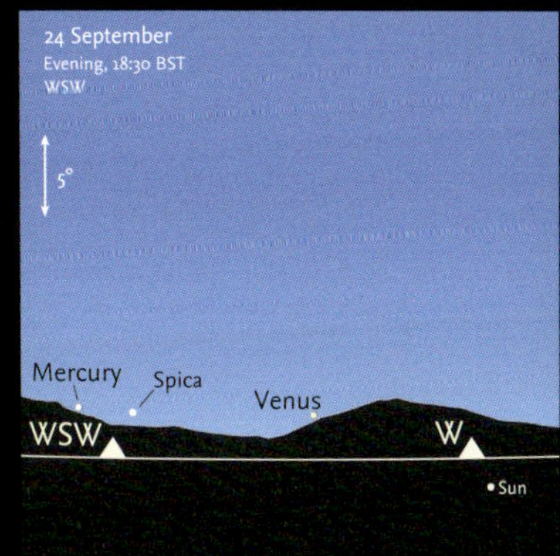

24 September • *Mercury at greatest eastern elongation with Venus to the west.*

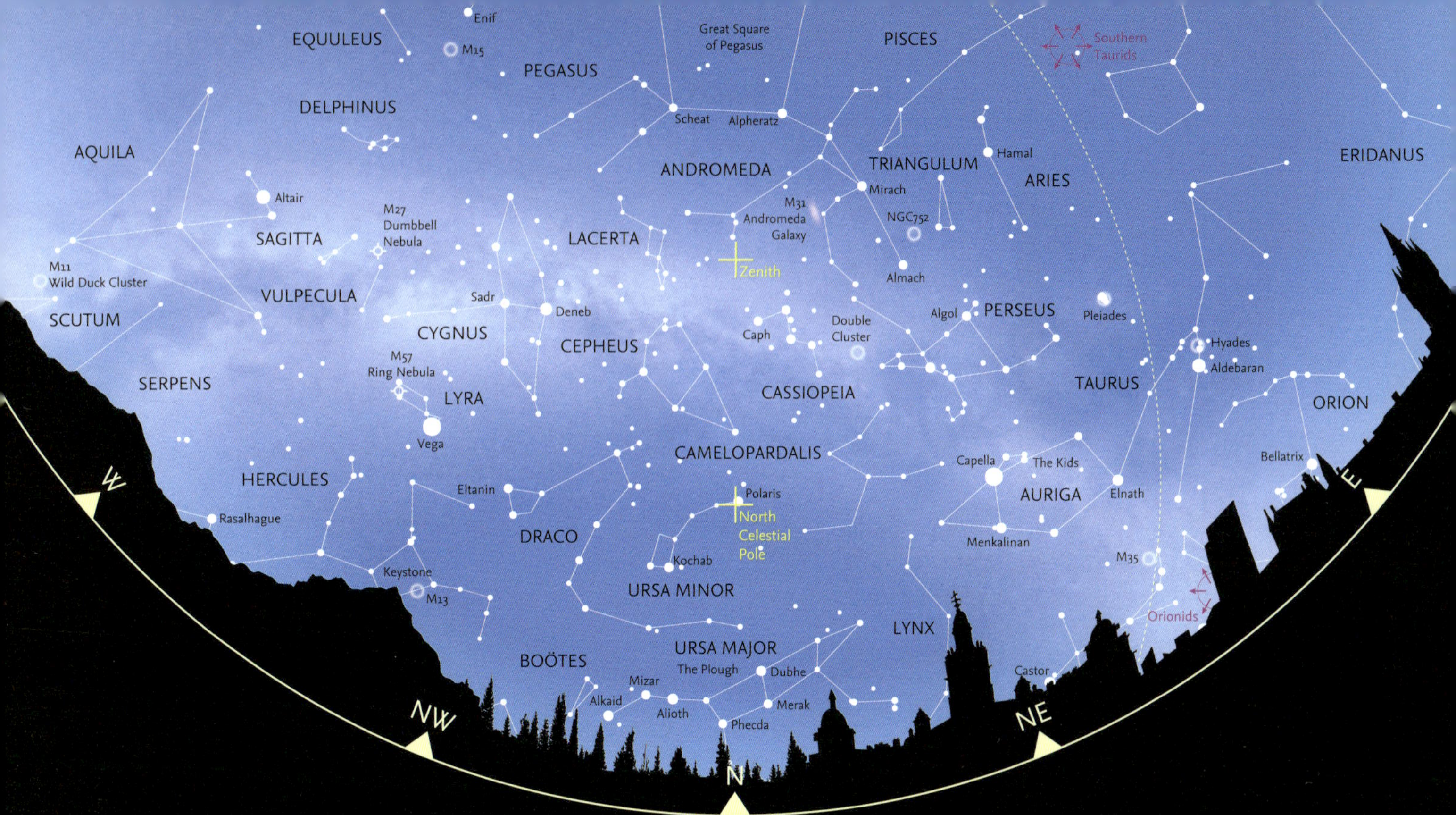

October – Looking North

1 October	00:00 BST (GMT + 1hr)
15 October	23:00 BST (GMT + 1hr)
1 November	21:00 GMT

October – Looking North

Ursa Major is grazing the horizon in the north, while high overhead are the constellations of ***Cepheus***, ***Cassiopeia*** and ***Perseus***, with the Milky Way between Cepheus and Cassiopeia near the zenith. ***Auriga*** is now clearly visible in the east, as is ***Taurus*** with the ***Pleiades***, ***Hyades*** and orange ***Aldebaran***. Also in the east, ***Orion*** and ***Gemini*** are starting to rise clear of the horizon.

The constellations of ***Boötes*** and ***Corona Borealis*** are now essentially lost to view in the northwest, and ***Hercules*** is also descending towards the western horizon. The three stars of the Summer Triangle are still clearly visible, although ***Aquila*** and ***Altair*** are beginning to approach the horizon in the west. Towards the end of the month (Sunday 31 October) Summer Time ends in Europe, with Britain reverting to Greenwich Mean Time and Europe to Central European Time.

Meteors

The ***Orionids*** meteor shower starts on 2 October and builds up to a broad maximum lasting around a week centred on 21–22 October. Like the May ***η-Aquariid*** shower, the Orionids are associated with Comet 1P/Halley. During this second pass through the stream of particles from the comet, slightly fewer meteors are seen than in May, but conditions are more favourable for northern observers. In both showers the meteors are very fast and many leave persistent trains, with hourly rates around 15. In 2027, there is a Last Quarter Moon lighting up the sky and interfering with observations after midnight.

The faint shower of the ***Southern Taurids*** (often with bright fireballs) continues through October, eventually reaching a fireball-dominant peak on 5 November. Towards the end of the month (around 20 October), another shower (the ***Northern Taurids)*** begins to show activity, peaking in November. The parent comet for both Taurid showers is Comet 2P/Encke. The meteors in both Taurid streams are relatively slow and bright.

A minor shower of short duration, the ***Draconids***, begins on 6 October and peaks on 9 October, when a waxing gibbous Moon will dominate the evening sky, setting just before midnight.

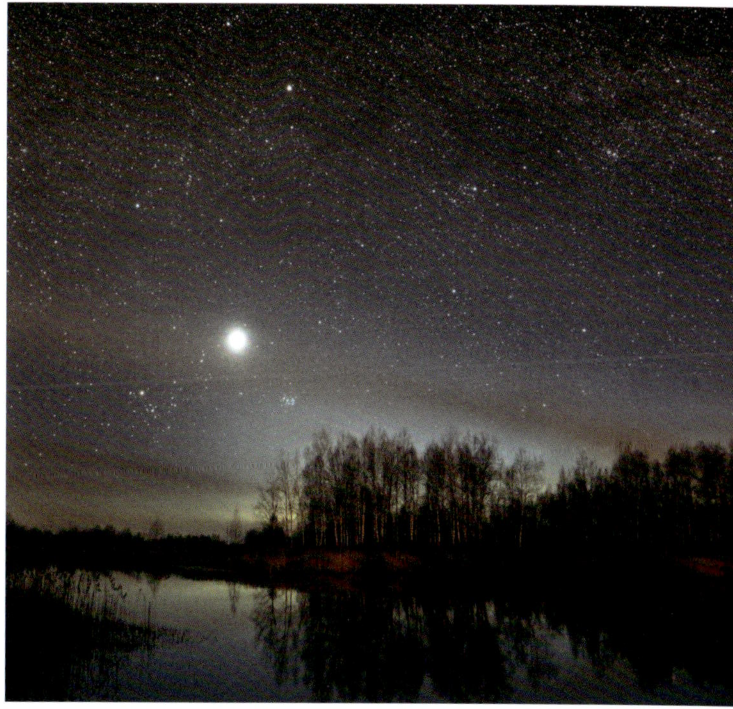

The brightest planet in the night sky, Venus, above the Hyades and Pleiades star clusters.

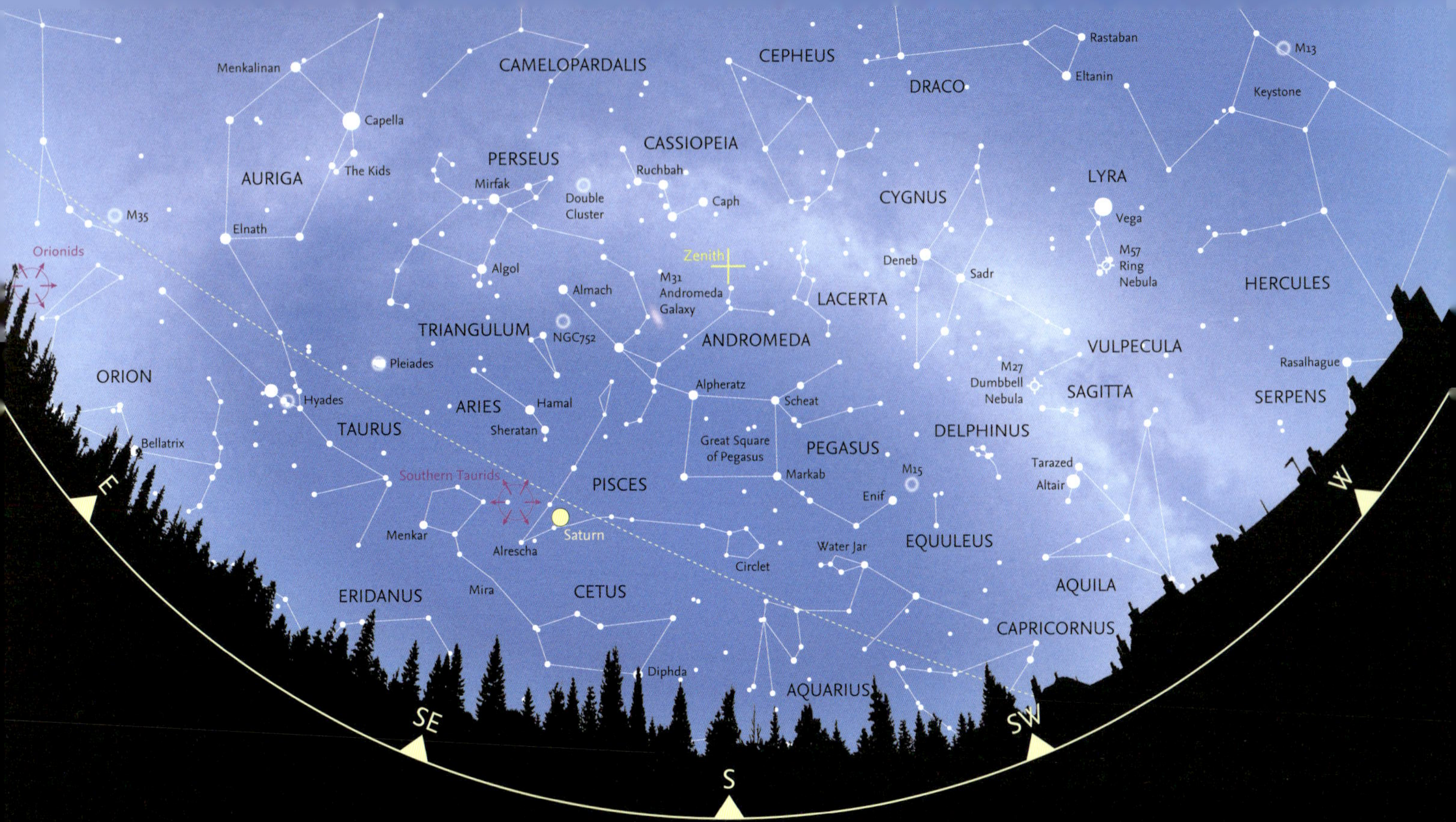

October – Looking South

1 October	00:00 BST (GMT + 1hr)
15 October	23:00 BST (GMT + 1hr)
1 November	21:00 GMT

October – Looking South

The Great Square of ***Pegasus*** dominates the southern sky, framed by the two chains of stars that form the constellation of ***Pisces***, together with ***Alrescha*** (α Piscium) at the point where the two lines of stars join. Also clearly visible is the constellation of ***Cetus***, below Pegasus and Pisces. Although ***Capricornus*** is beginning to disappear, ***Aquarius*** to its east is well placed in the south, with solitary ***Fomalhaut*** and the constellation of ***Piscis Austrinus*** (the Southern Fish) beneath it, close to the horizon.

The main band of the Milky Way and the Great Dark Rift runs down from ***Cygnus***, through ***Vulpecula***, ***Sagitta*** and ***Aquila*** towards the western horizon. ***Delphinus*** and the tiny, unremarkable constellation of ***Equuleus*** lie between the band of the Milky Way and Pegasus. ***Andromeda*** is clearly visible high in the sky to the southeast, with the small constellation of ***Triangulum*** and the zodiacal constellation of ***Aries*** below it. ***Perseus*** is high in the east, and by now the ***Pleiades*** and ***Taurus*** are well clear of the horizon. Later in the night, and later in the month, ***Orion*** rises in the east, a sign that the autumn season has arrived, and winter is approaching.

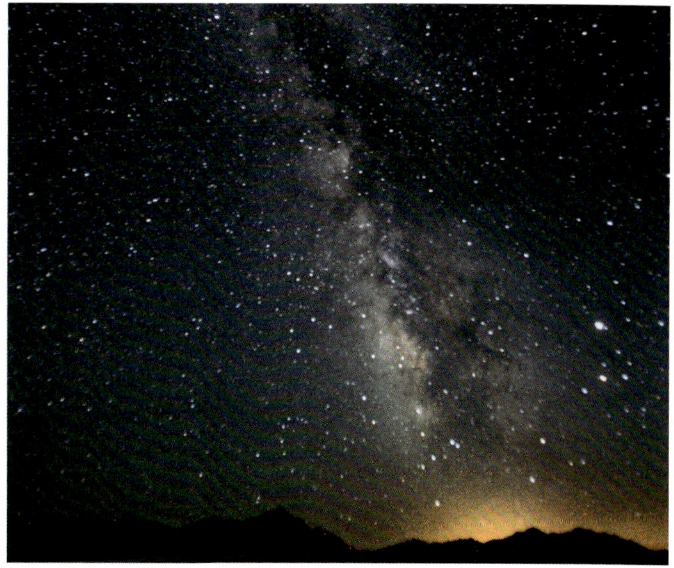

The Milky Way and the Great Dark Rift, caused by thick dust.

The Moon's phases for October 2027

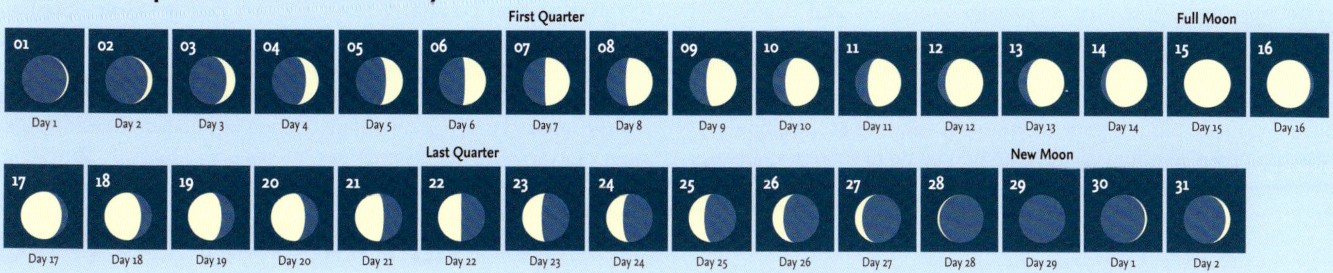

October – Moon and Planets

The Moon

The waxing crescent Moon lies 3.1°S of ***Spica*** on the first day of October, both lying very low on the horizon after sunset. On 2 October ***Mercury*** (mag. 0.3) sits 1.8°N of the Moon and the following day ***Mars*** (mag. 1.4) lies 4.5°N of the Moon, both planets visible only at dusk. On 4 October ***Antares*** will be 0.2°S of the Moon in the early evening sky. On 7 October the Moon arrives at First Quarter, on 15 October it becomes Full. On 18 October the waning gibbous Moon passes 0.6°N of the ***Pleiades***. On 22 October a Last Quarter Moon dominates the dawn sky, two days later the waning crescent Moon passes 2.6°S of ***Regulus*** in ***Leo***, above the horizon a few hours after midnight. On 26 October ***Jupiter*** (mag. -1.8) can be found 4.9°N of the Moon before sunrise. The lunar cycle restarts on 29 October with a New Moon and the month ends with ***Venus*** (mag. -3.9) 4.4°N of the Moon, visible just after sunset.

The Planets

Mercury continues to be seen in the early evening sky, then in the early morning from 19 October. It dims dramatically at the solar conjunction and bounces back to mag. -0.1 at the end of October. It appears at mag. 1.2 3.8°S of Venus (mag. -3.9) on 11 October. ***Venus*** stays in the early evening sky at mag. -3.9. ***Mars*** moves from ***Libra*** to ***Scorpio*** from 15 October, then shifts into ***Ophiuchus*** from 27 October, low on the horizon just after sunset all month. On 29 October it appears 3.7°N of ***Antares***. ***Jupiter*** continues in ***Leo***, visible now a few hours before sunrise, brightening from mag. -1.7 to -1.8. ***Saturn*** lies in ***Pisces***, climbing above the horizon late in the evening at mag. 0.2. It reaches opposition on 18 October, brightening to mag. 0.1 before fading back to mag. 0.2 at the end of the month. ***Uranus*** stays in ***Taurus***, above the horizon later in the evening at mag. 5.6. ***Neptune*** is in Pisces, visible from a few hours after sunset, at mag. 7.7.

The path of the Sun and the planets along the ecliptic in October.

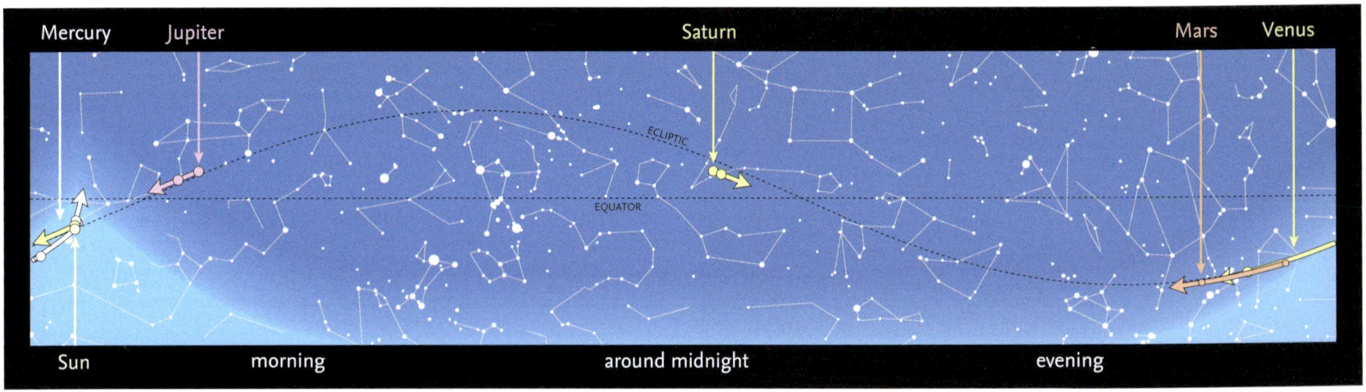

Calendar for October

01	09:12	Spica 3.1°N of the Moon
02	00:27	Mercury (mag. 0.3) 1.8°N of the Moon
03	08:58	Mars (mag. 1.4) 4.5°N of the Moon
04	17:01	Antares 0.2°S of the Moon
07	11:47	First Quarter Moon
09	15:47	Moon at apogee = 404,531 km
11	14:00	Mercury (mag. 1.2) 3.8°S of Venus (mag. -3.9)
15	13:47	Full Moon
18	01:27	Saturn (mag. 0.1) at opposition
18	10:48	Pleiades 0.6°S of the Moon
21–22		Orionid meteor shower maximum
22	00:00	Last Quarter Moon
24	23:09	Regulus 2.6°N of the Moon
25	05:36	Moon at perigee = 368,651 km
26	09:33	Jupiter (mag. -1.8) 4.9°N of the Moon
29	13:36	New Moon
29	13:47	Mars (mag. 1.4) 3.7°N of Antares

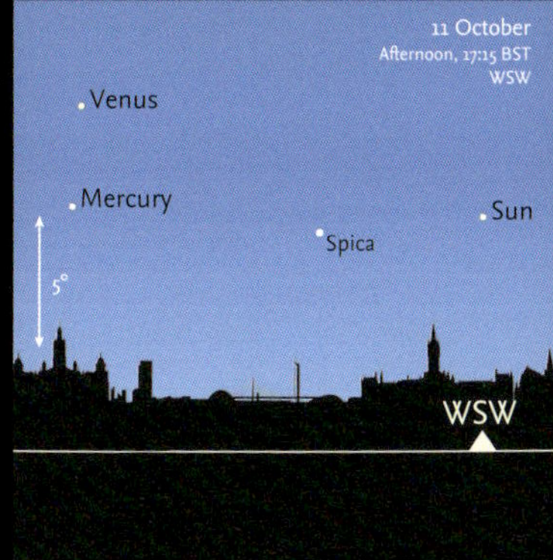

11 October • *Mercury and Venus setting with the Sun.*

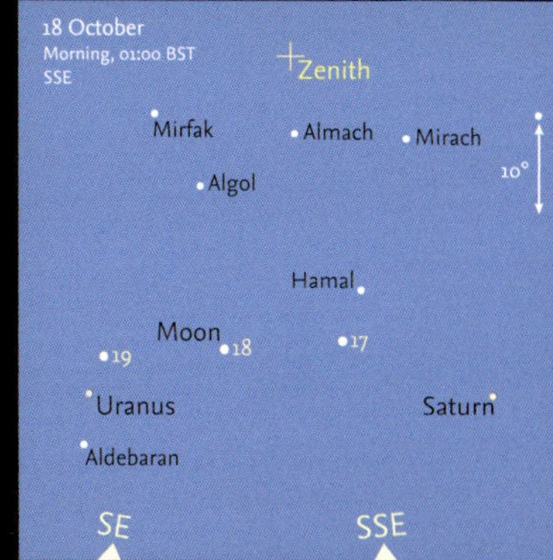

18 October • *Saturn at opposition. The waning gibbous Moon, Uranus and Aldebaran lie to the east.*

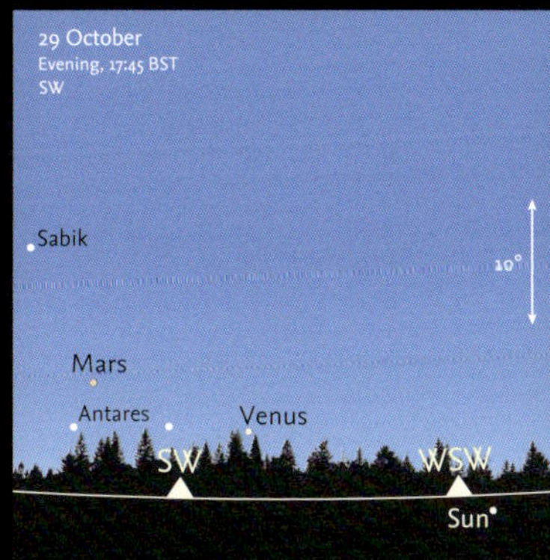

29 October • *Mars and its red rival, Antares.*

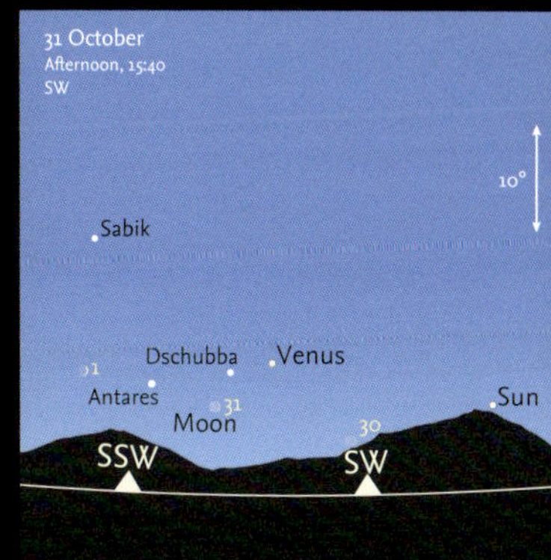

31 October • *Venus close to the thin waxing crescent Moon.*

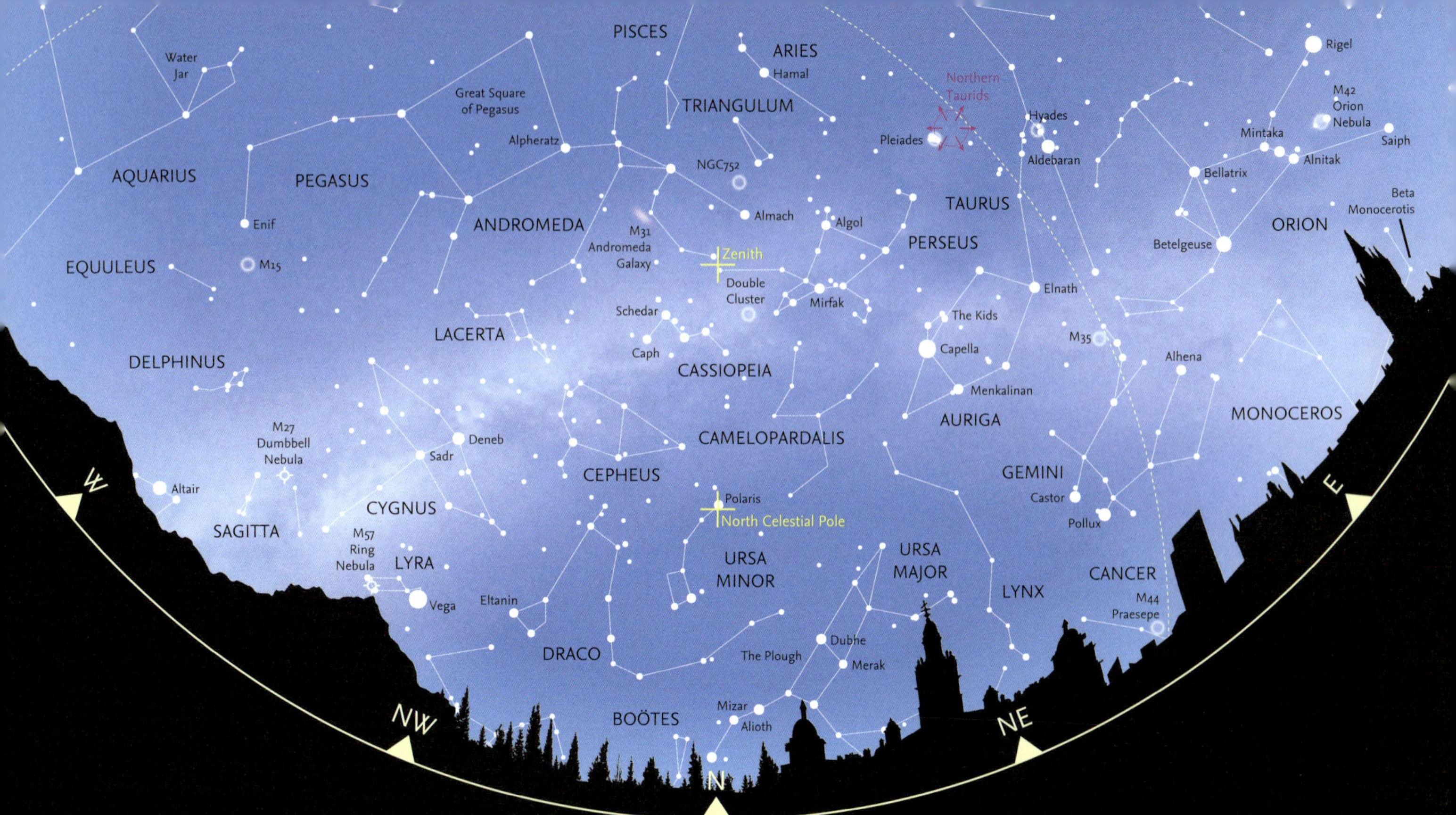

November – Looking North

1 November	23:00 GMT
15 November	22:00 GMT
1 December	21:00 GMT

November – Looking North

Most of ***Aquila*** has now disappeared below the horizon, but two of the stars of the Summer Triangle, ***Vega*** in ***Lyra*** and ***Deneb*** in ***Cygnus***, are still clearly visible in the west. The head of ***Draco*** is now low in the northwest and only a small portion of ***Hercules*** remains above the horizon. The southernmost stars of ***Ursa Major*** are now coming into view. The Milky Way arches overhead, with the denser star clouds in the west and the less heavily populated region through ***Auriga*** and ***Monoceros*** in the east. The star β ***Monocerotis*** is actually a triple star system; all three stars are six to seven times the mass of the Sun. High overhead, ***Cassiopeia*** is near the zenith and ***Cepheus*** has swung round to the northwest, while Auriga is now high in the northeast. ***Gemini***, with ***Castor*** and ***Pollux***, is well clear of the eastern horizon, and ***Procyon*** (α Canis Minoris), in reality a binary star system, is just climbing into view almost due east.

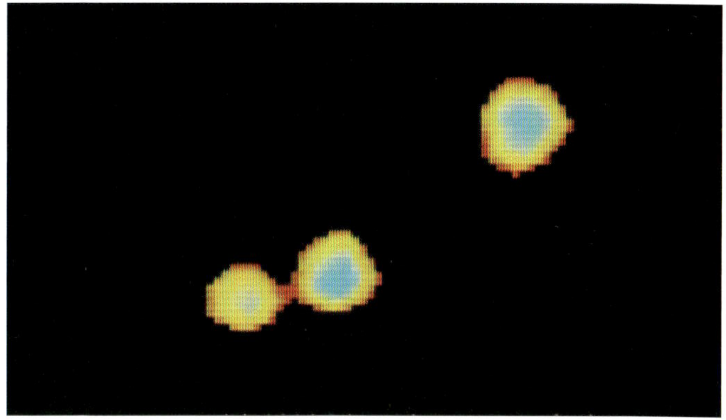

The triple star system Beta Monocerotis.

Meteors

The ***Southern Taurid*** shower of fireball meteors peaks on 5 November and the ***Northern Taurids***, which began in mid-October, reaches maximum – although only with a rate of about five meteors per hour – on 12–13 November. The Moon is close to First Quarter on 5 November and sets a few hours before midnight, it's even brighter in its waxing gibbous phase on 12 and 13 November and will dominate the sky throughout the night. Far more striking, however, are the ***Leonids***, which have a relatively short period of activity (6–30 November), with maximum on the evening of 17 November and early morning of 18 November. This shower is associated with Comet 55P/Tempel-Tuttle and has shown extraordinary activity on various occasions with many thousands of meteors per hour. The rate is likely to be about 10–15 per hour at peak activity; however, the Moon will be waning gibbous and above the horizon all night. These meteors are the fastest shower meteors recorded (about 70 km per second) and often leave persistent trains.

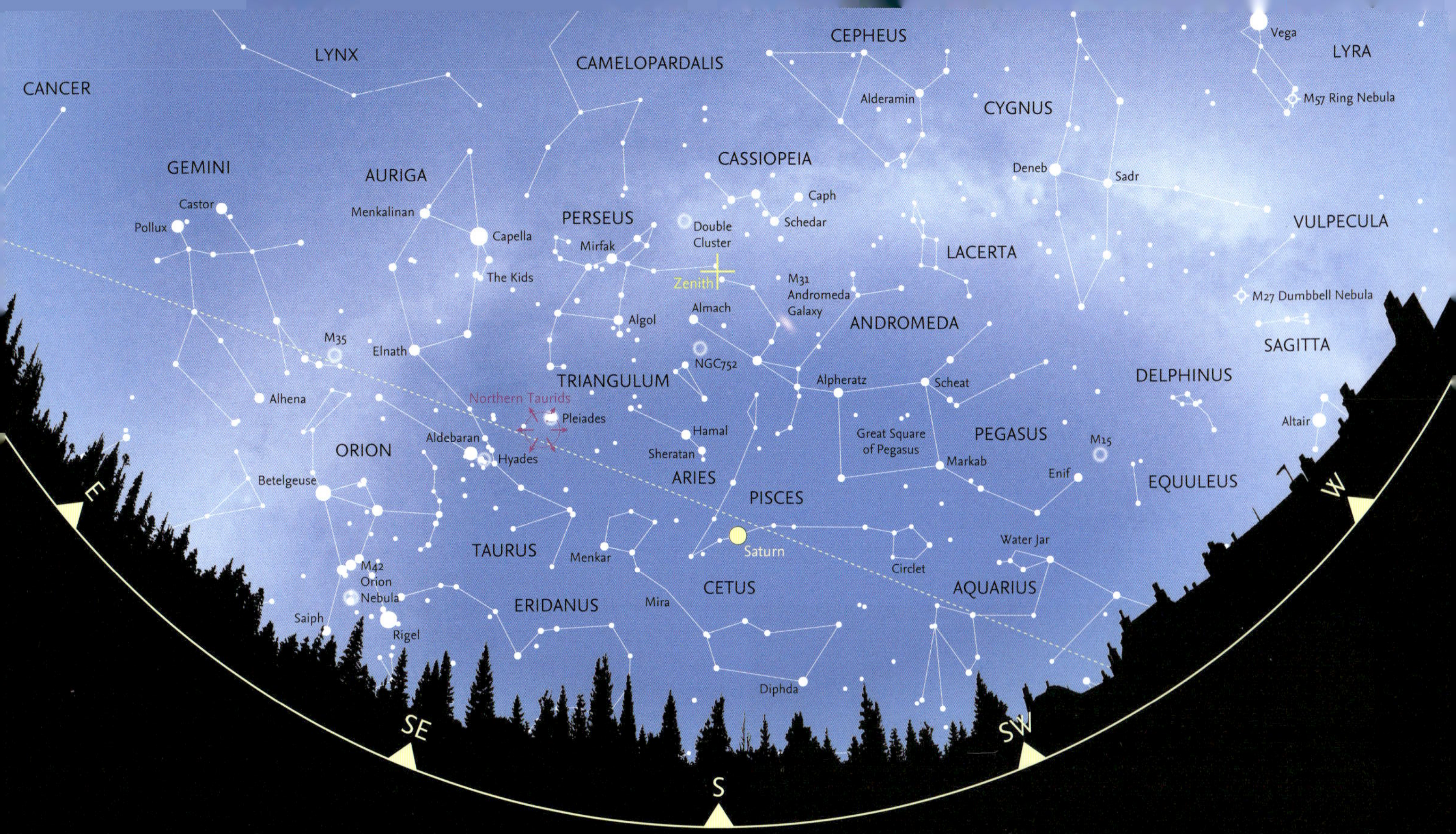

November – Looking South

1 November	23:00 GMT
15 November	22:00 GMT
1 December	21:00 GMT

November – Looking South

Orion has now risen above the eastern horizon, and part of the long, straggling constellation of ***Eridanus*** (which begins near ***Rigel***) is visible to the west of Orion. Higher in the sky, ***Taurus***, with the ***Pleiades*** cluster, and orange ***Aldebaran*** are now easy to observe. To their west, both ***Pisces*** and ***Cetus*** are close to the meridian. The famous long-period variable star ***Mira*** (ο Ceti), with a typical range of magnitude 3.4 to 9.5, is favourably placed for observation. In the southwest, ***Capricornus*** has slipped below the horizon, but ***Aquarius*** remains visible. Even farther west, ***Altair*** may be seen early in the night, but most of ***Aquila*** has already disappeared from view. ***Delphinus***, together with ***Sagitta*** and ***Vulpecula*** in the Milky Way, will soon vanish for another year. Both ***Pegasus*** and ***Andromeda*** are easy to see, and one of the lines of stars that make up Andromeda finishes close to the zenith, which is also close to one of the outlying stars of ***Perseus***, high in the east.

The constellation Pisces.

The Moon's phases for November 2027

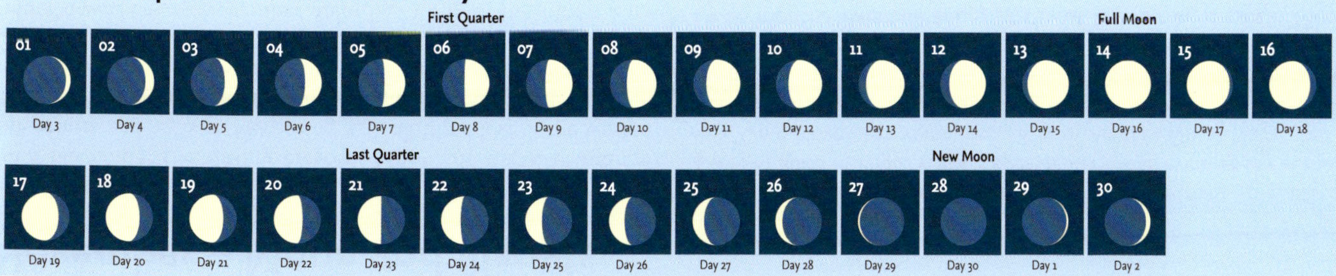

November – Moon and Planets

The Moon

On 1 November the waxing crescent Moon sits 0.4°N of orange-red ***Antares*** and 3.3°S of ***Mars*** (mag. 1.3), all three low in the western sky at sunset. On 6 November there will be a First Quarter Moon, it becomes Full on 14 November and passes 0.5°N of the ***Pleiades***. The Last Quarter Moon arrives on 21 November and passes 2.9°S of ***Regulus*** before dawn. On 25 November ***Spica*** lies 3.2°N of the Moon a few hours before sunrise, three days later the Moon is New. The end of the month is marked by two planets sitting close to the waxing crescent Moon at dusk: ***Mars*** (mag. 1.3) lies 1.6°N and ***Venus*** (mag. -3.9) will be 0.9°N of the Moon.

The Planets

Mercury is visible at dawn, it climbs away from the Sun to western elongation on 4 November. It brightens from mag. -0.2 to -0.9 over the course of the month. On 5 November Mercury (mag. -0.6) passes 4.0°N of ***Spica*** in ***Virgo***. ***Venus*** is spectacular in the evening sky at mag. -3.9. On 9 November Venus lies 3.8°N of ***Antares***. On 25 November it sits 0.3°S of ***Mars*** (mag. 1.3) and on 30 November Venus passes 0.9°N of the waxing crescent Moon. Mars is in ***Ophiuchus*** and moves into ***Sagittarius*** from 21 November, low in the sky just after sunset at mag. 1.3 all month. Mars will appear 3.3°N of the waxing crescent Moon on the first day of the month and it approaches the waxing crescent Moon again on the last day of November, at 1.6°N. ***Jupiter*** moves from its long stay in ***Leo*** to Virgo on 23 November, appearing above the horizon several hours after midnight, brightening from mag. -1.8 to -1.9. ***Saturn*** lies in ***Pisces***, visible from sunset, fading slightly from mag. 0.2 to 0.3. ***Uranus*** stays in ***Taurus***, above the horizon a few hours after sunset, at mag. 5.6. It reaches opposition on 30 November. ***Neptune*** is in Pisces visible from sunset, at mag. 7.7.

The path of the Sun and the planets along the ecliptic in November.

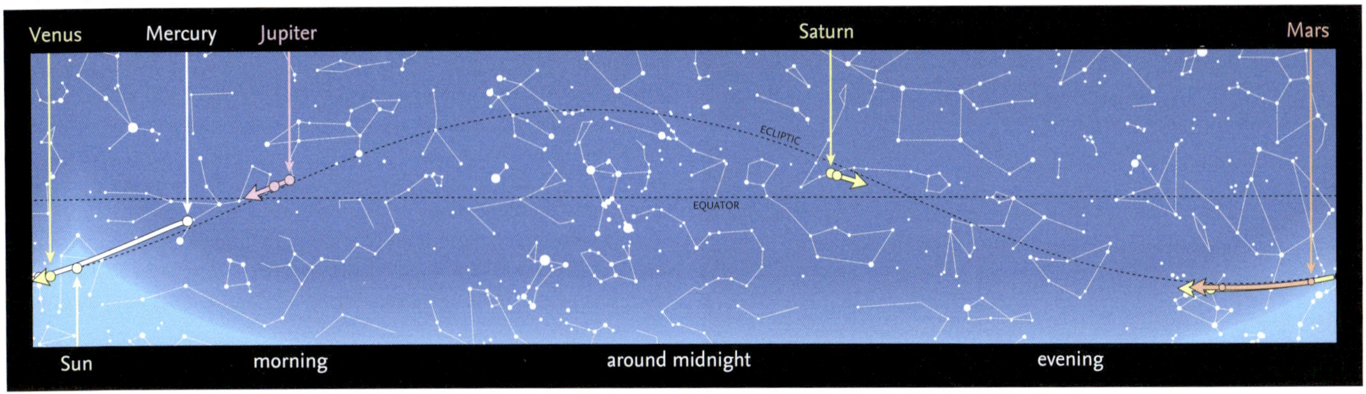

Calendar for November

01	02:13	Antares 0.4°S of the Moon
01	06:53	Mars (mag. 1.3) 3.3°N of the Moon
04	07:58	Mercury at greatest elongation (18.8°W, mag. -0.5)
05	00:48	Mercury (mag. -0.6) 4.0°N of Spica
05		Southern Taurid meteor shower maximum
06	08:00	First Quarter Moon
06	11:37	Moon at apogee = 404,183 km
09	06:44	Venus (mag. -3.9) 3.8°N of Antares
12–13		Northern Taurid meteor shower maximum
14	03:26	Full Moon
14	18:14	Pleiades 0.5°S of the Moon
17–18		Leonid meteor shower maximum
19	00:10	Moon at perigee = 369,356 km
21	00:48	Last Quarter Moon
21	04:52	Regulus 2.9°N of the Moon
25	01:00	Venus (mag. -3.9) 0.3°S of Mars (mag. 1.3)
25	02:24	Spica 3.2°N of the Moon
28	03:24	New Moon
30	07:53	Mars (mag. 1.3) 1.6°N of the Moon
30	09:13	Uranus (mag. 5.6) at opposition
30	13:20	Venus (mag. -3.9) 0.9°N of the Moon

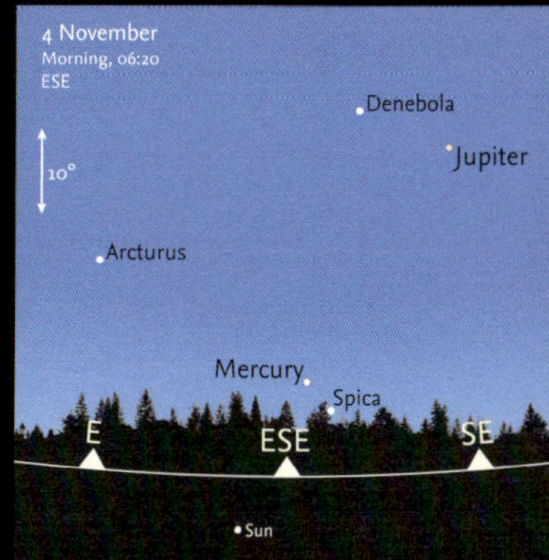

4 November • *Mercury at greatest western elongation. Arcturus and Jupiter sit nearby.*

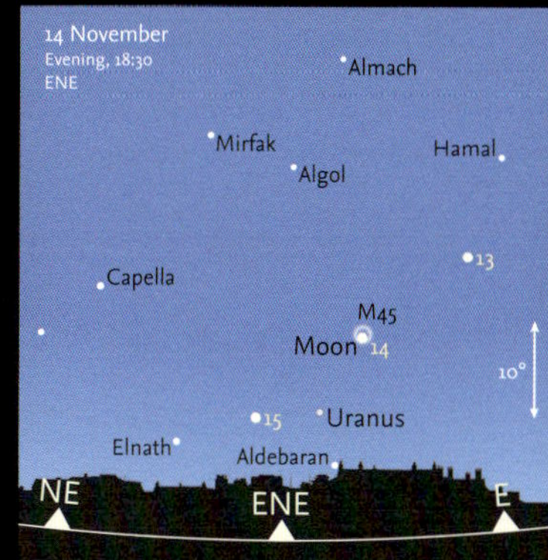

14 November • *The Pleiades (M45) and the Full Moon together, with Uranus and Aldebaran close by.*

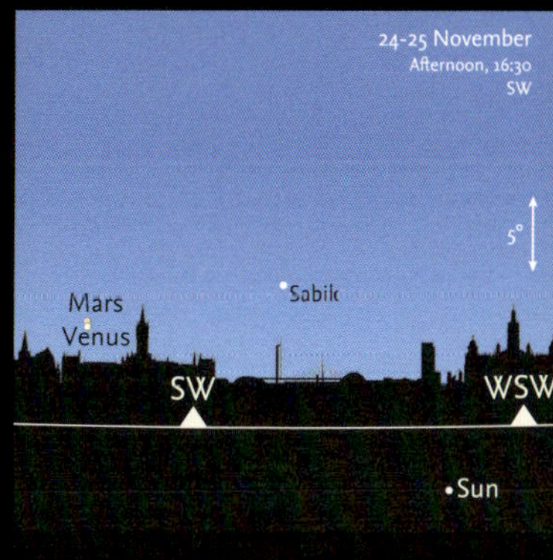

24–25 November • *Venus and Mars sit next to each other.*

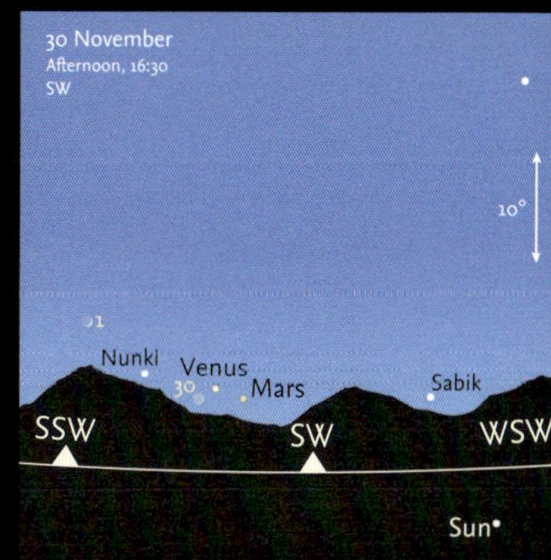

30 November • *Venus and Mars by the thin waxing crescent Moon.*

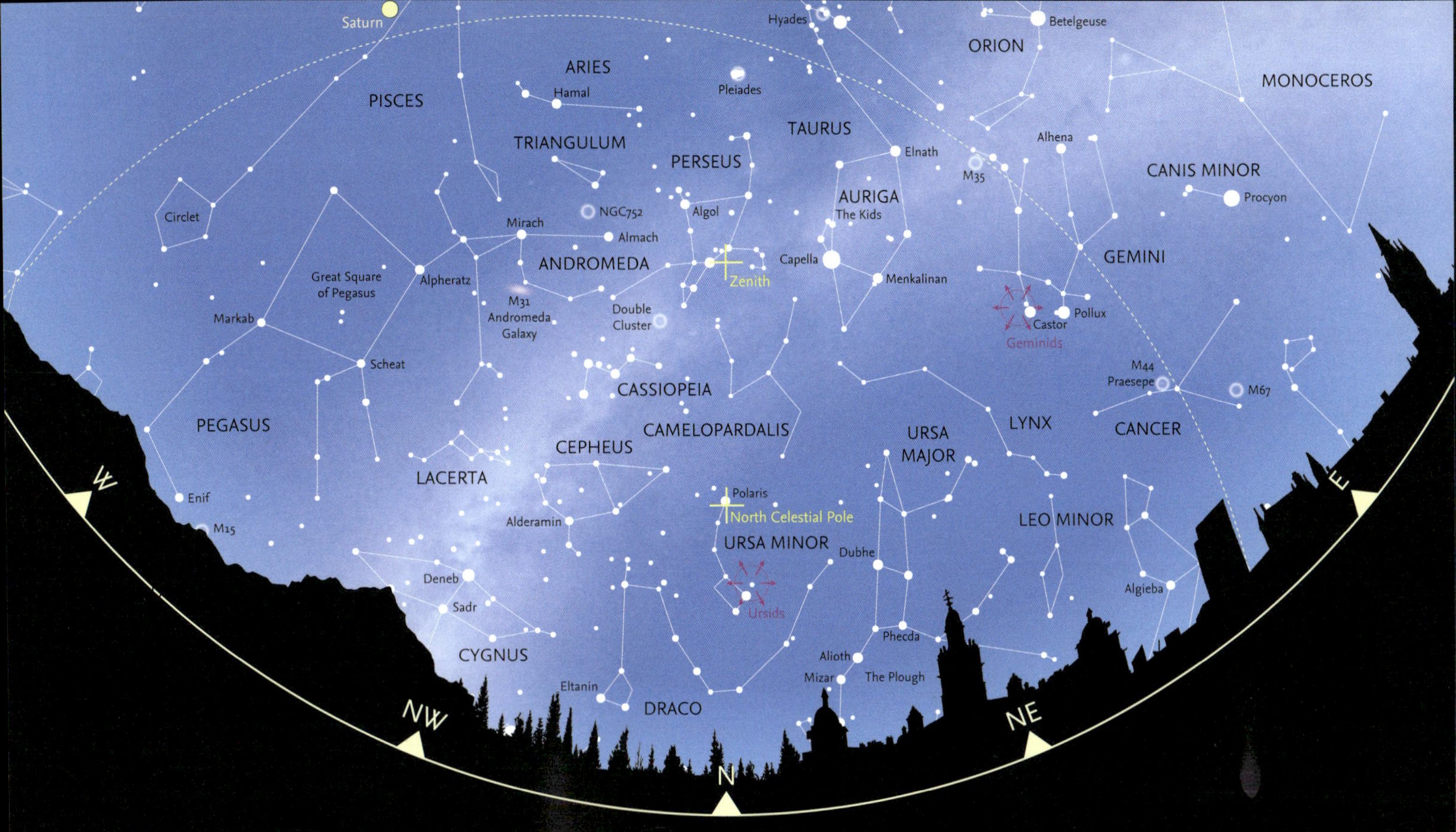

December – Looking North

1 December	23:00 GMT
15 December	22:00 GMT
1 January	21:00 GMT

December – Looking North

The winter solstice on 22 December brings the shortest period of daylight.

The Great Bear, ***Ursa Major*** has now swung around and is starting to 'climb' in the east. The fainter stars in the southern part of the constellation are now fully in view. The other bear, ***Ursa Minor***, 'hangs' below ***Polaris*** in the north. Directly above it is the faint constellation of ***Camelopardalis***, with the other circumpolar constellation, ***Lynx***, to its east. ***Vega*** (α Lyrae) is skimming the horizon in the northwest, but ***Deneb*** (α Cygni) and most of ***Cygnus*** remain visible farther west. In the east, ***Regulus*** (α Leonis) and the constellation of ***Leo*** are beginning to rise above the horizon. ***Cancer*** stands high in the east, with ***Gemini*** even higher in the sky. ***Perseus*** is at the zenith, with ***Auriga*** and ***Capella*** between it and Gemini. Because it is so high in the sky, now is a good time to examine the star clouds of the fainter portion of the Milky Way, between ***Cassiopeia*** in the west to Gemini and ***Orion*** in the east.

Meteors

There is one significant meteor shower in December (the last major shower of the year). This is the ***Geminid*** shower, which is visible over the period 4–17 December and comes to maximum on 14 December, one day after the Full Moon, so significant moonlight will affect observing conditions. It is one of the most active showers of the year, with a possible rate of around 100 meteors per hour. It is the one major shower that shows good activity before midnight. The source of the Geminids is debris from an asteroid called 3200 Phaethon; most meteor showers originate from cometary debris. The Geminids are assumed to consist of denser, rocky material, they are slower than most other meteors and often appear to last longer. The brightest often break up into numerous luminous fragments that follow similar paths across the sky. There is a second shower: the ***Ursids***, active in the last half of the month and peaking on 22–23 December, with a rate at maximum of 5–10, occasionally rising to 25 per hour. The peak in 2027 occurs when the Moon is a waning crescent rising a few hours after midnight, so conditions are favourable for most of the night. The parent body is Comet 8P/Tuttle.

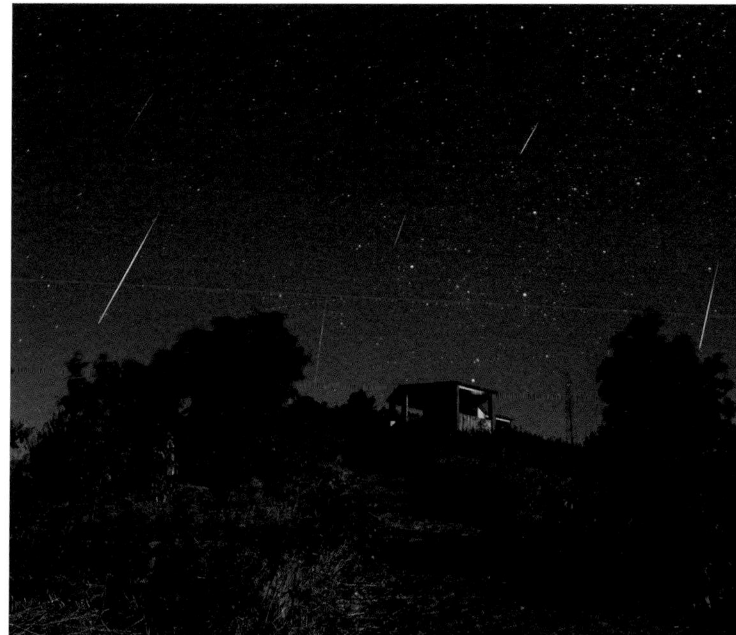

The winter Geminid meteor shower.

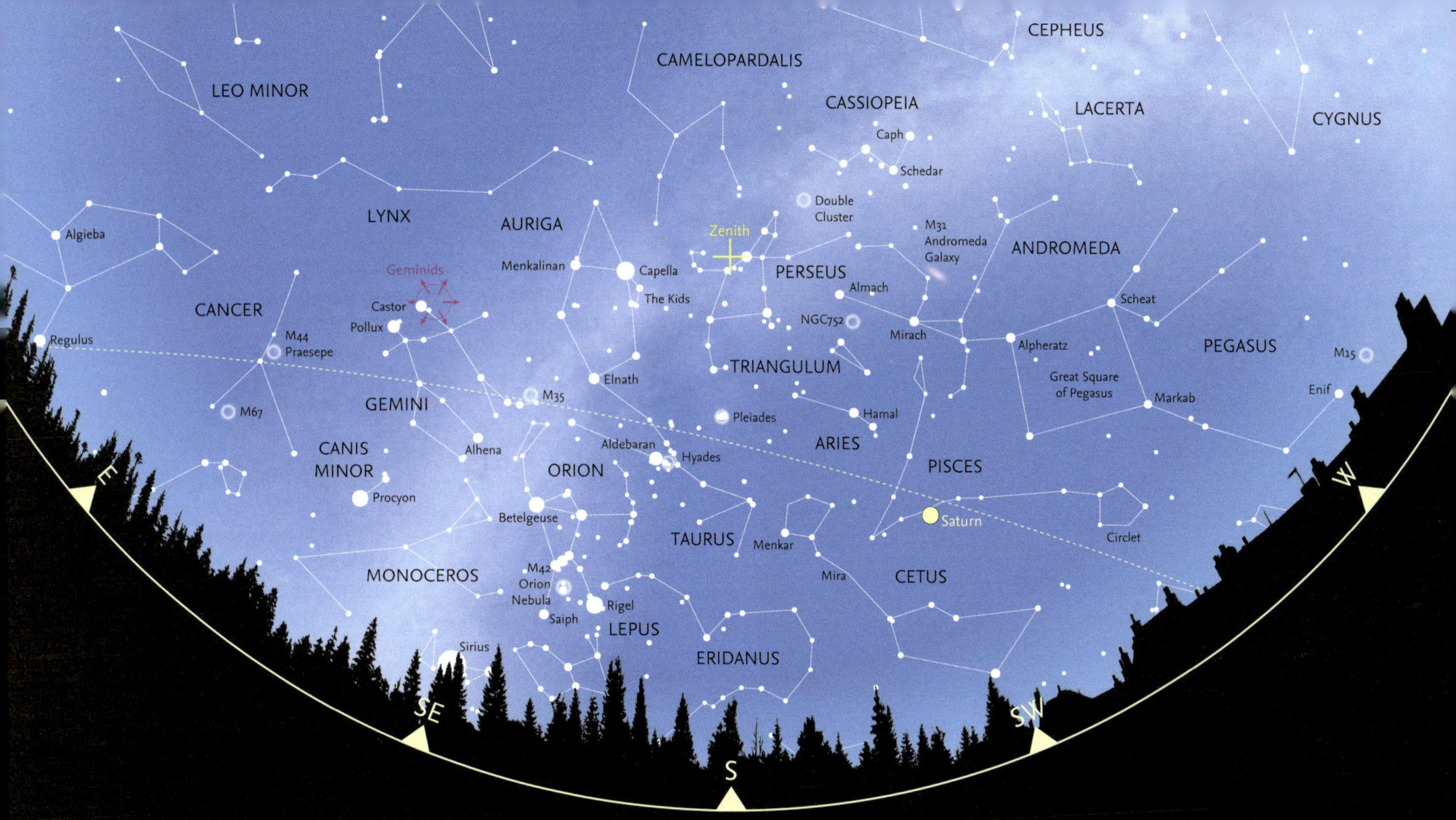

December – Looking South

1 December	23:00 GMT
15 December	22:00 GMT
1 January	21:00 GMT

December – Looking South

The fine open cluster of the ***Pleiades*** is due south around 22:00, with the ***Hyades*** cluster, ***Aldebaran*** and the rest of ***Taurus*** clearly visible to the east. ***Auriga*** (with ***Capella***) and ***Gemini*** (with ***Castor*** and ***Pollux***) are both well placed for observation. ***Orion*** has made a welcome return to the winter sky, and both ***Canis Minor*** (with ***Procyon***) and ***Canis Major*** (with ***Sirius***, the brightest star in the sky) are starting to rise later in the evening. The small, poorly known constellation of ***Lepus*** lies to the south of Orion. In the west, ***Aquarius*** has now disappeared, and ***Cetus*** is becoming lower, but ***Pisces*** is still easily seen, as are the constellations of ***Aries***, ***Triangulum*** and ***Andromeda*** above it. The Great Square of ***Pegasus*** is starting to plunge down towards the western horizon and, because of its orientation on the sky, appears more like a large diamond standing on one point than a square.

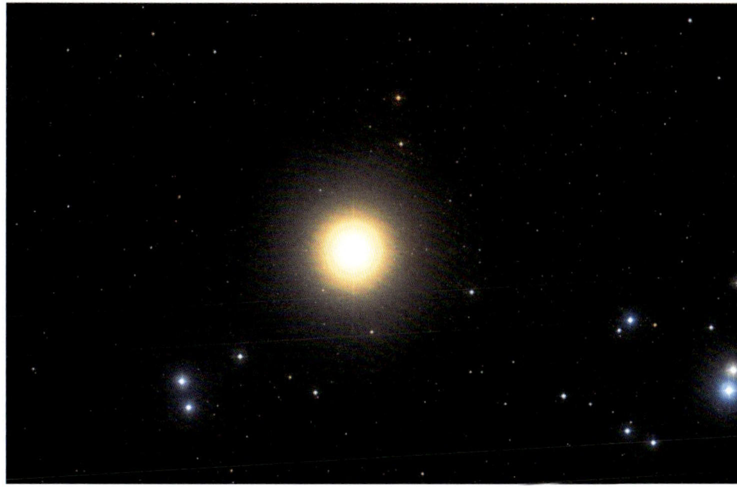

The red supergiant Betelgeuse.

The Moon's phases for December 2027

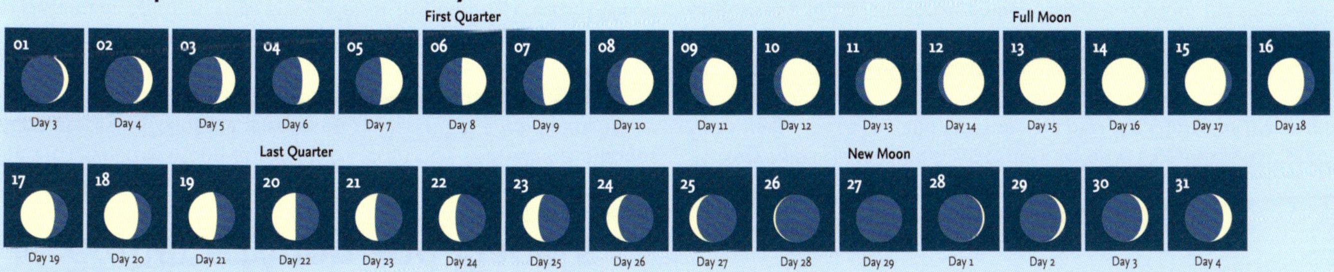

December – Moon and Planets

The Moon

The First Quarter Moon arises on 6 December. The waxing gibbous Moon passes 0.5°N of the ***Pleiades*** on 12 December a few hours before dawn. On 13 December the Moon becomes Full. ***Regulus*** lies 3.1°N of the waning gibbous Moon on 18 December, best seen a few hours before sunrise. Two days later the Moon arrives at Last Quarter. On 22 December ***Spica*** in ***Virgo*** sits 3.3°N of the waning crescent Moon before dawn. On 25 December ***Antares*** can be found 0.4°S of the Moon visible just before sunrise and two days later the lunar phase cycle resets with a New Moon. On 29 December ***Mars*** (mag. 1.2) passes 0.5°S of the Moon and on the last day of December 2027 ***Venus*** (mag. -4.0) will be found 3.1°S of the Moon. Both events can be observed at sunset.

The Planets

Mercury switches from the dawn to dusk sky on 11 December. It brightens from mag. -0.1 to -1.3 and fades to -0.9 over the course of the month. ***Venus*** is beautiful in the evening sky, brightening from mag. -3.9 to -4.0. Venus appears 3.1°S of the waxing crescent Moon on 31 December. ***Mars*** is in ***Sagittarius*** and sets with the Sun; it starts the month at mag. 1.3, brightening to 1.2. On 29 December Mars sits 0.5°S of the thin waxing crescent Moon. ***Jupiter*** is in ***Virgo*** and pops above the horizon an hour after midnight, brightening from mag. -1.9 to -2.1. ***Saturn*** lies in ***Pisces***, visible from sunset, dimming over the course of the month from mag. 0.3 to 0.5. It ends its retrograde motion on 24 December. ***Uranus*** stays in ***Taurus***, visible shortly after sunset, at mag. 5.6. ***Neptune*** is in Pisces visible from sunset, starting at mag. 7.7 and fading a little to mag. 7.8. It reverts back to direct motion on 15 December.

The path of the Sun and the planets along the ecliptic in December.

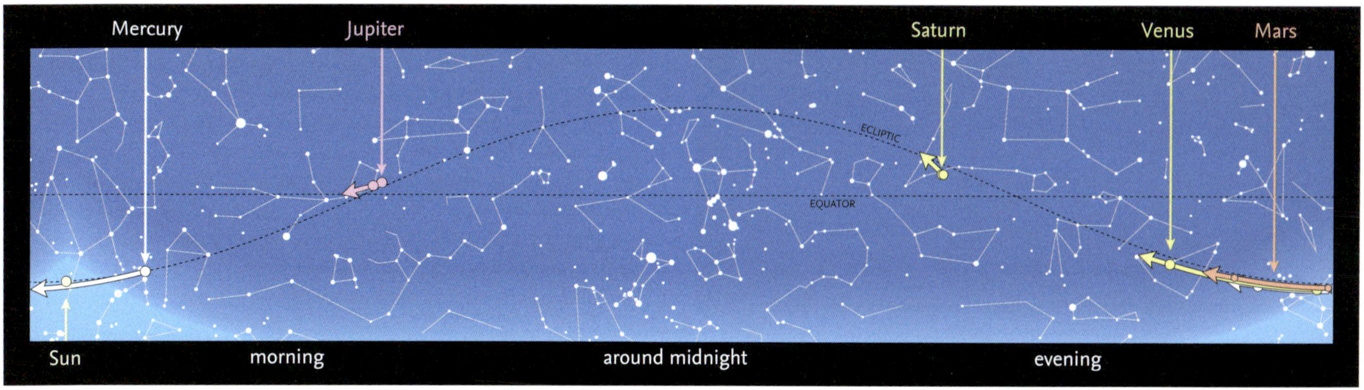

Calendar for December

04	08:42	Moon at apogee = 404,663 km
06	05:22	First Quarter Moon
12	03:55	Pleiades 0.5°S of the Moon
13	16:09	Full Moon
14		Geminid meteor shower maximum
15	11:13	Neptune ends retrograde motion
16	02:21	Moon at perigee = 364,026 km
18	10:50	Regulus 3.1°N of the Moon
20	09:11	Last Quarter Moon
22	02:43	Winter Solstice
22	08:13	Spica 3.3°N of the Moon
22–23		Ursid meteor shower maximum
24	03:04	Saturn ends retrograde motion
25	18:01	Antares 0.4°S of the Moon
27	20:12	New Moon
29	12:10	Mars (mag. 1.2) 0.5°S of the Moon
31	01:54	Venus (mag. -4.0) 3.1°S of the Moon

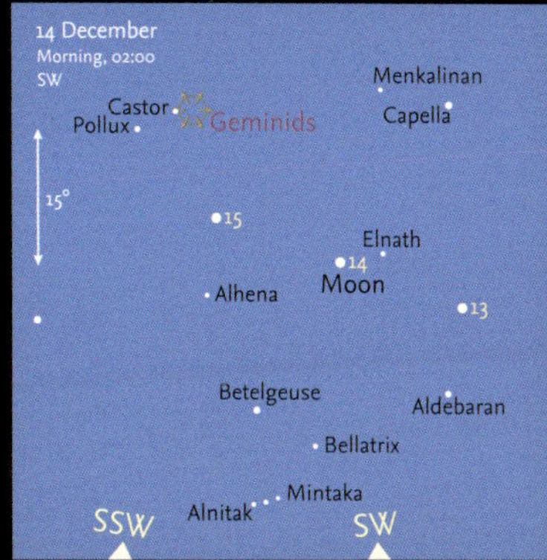

14 December • *The Geminid meteor shower peaks, and moonlight interferes with observations.*

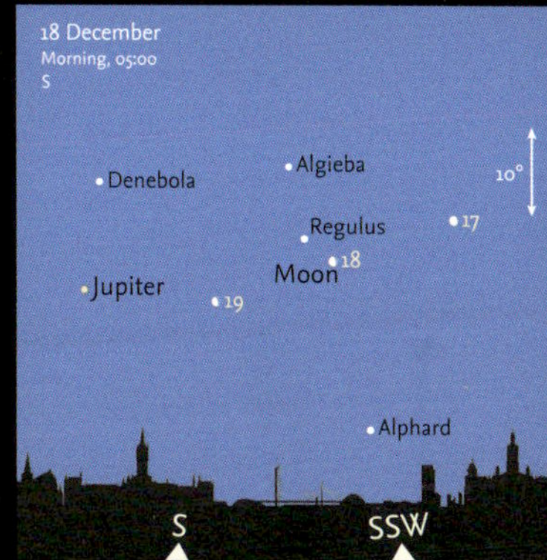

18 December • *Regulus and the waning gibbous Moon.*

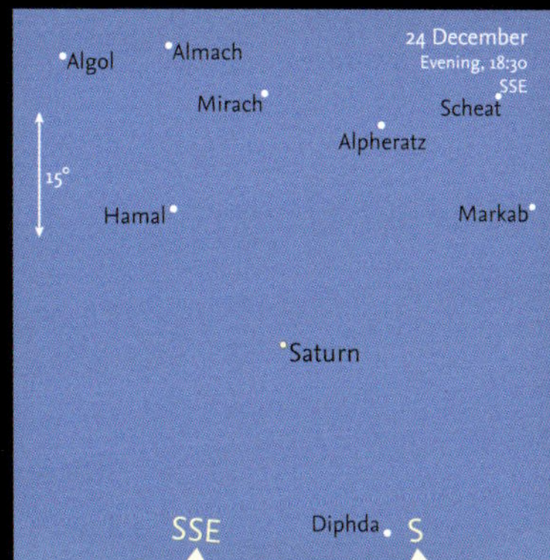

24 December • *Saturn ends retrograde motion.*

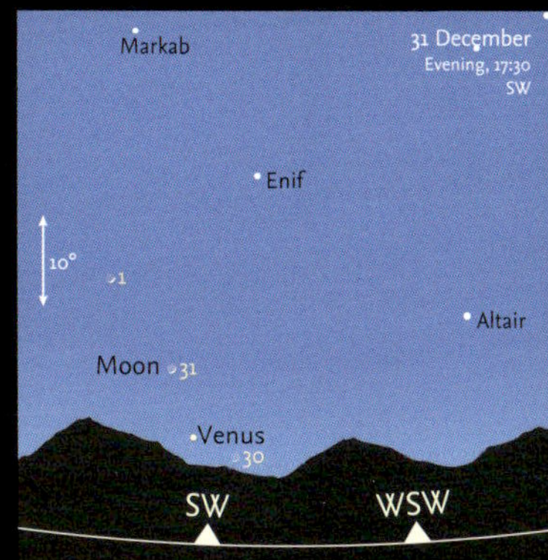

31 December • *Venus and the waxing crescent Moon together. Altair lies to the west.*

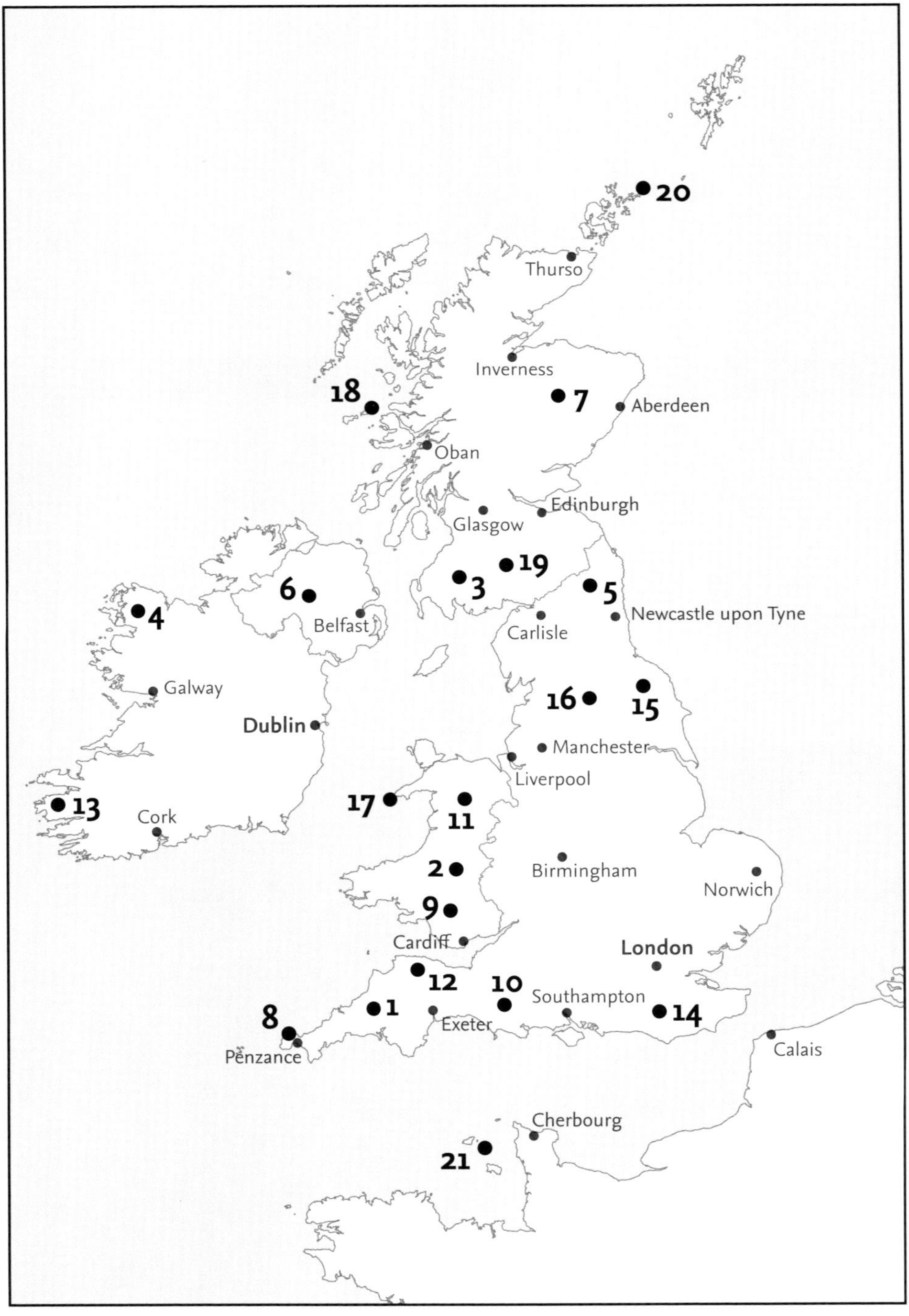
20
Thurso
Inverness
18
7
Aberdeen
Oban
Edinburgh
Glasgow
19
6
3
5
4
Belfast
Newcastle upon Tyne
Carlisle
Galway
16
15
Dublin
Manchester
Liverpool
17
11
13
Cork
2
Birmingham
Norwich
9
Cardiff
London
12
10
Southampton
8
1
Exeter
14
Penzance
Calais
Cherbourg
21

Dark Sky Sites

International Dark-Sky Association Sites

The ***International Dark-Sky Association*** (IDA) recognizes various categories of sites that offer areas where the sky is dark at night, free from light pollution and particularly suitable for astronomical observing. A number of sites in Great Britain and Ireland have been given specific recognition and are shown on the map. These are:

Parks

1 ***Bodmin Moor Dark Sky Landscape***
2 ***Elan Valley Estate***
3 ***Galloway Forest Park***
4 ***Mayo Dark Sky Park***
5 ***Northumberland National Park and Kielder Water & Forest Park***
6 ***OM Dark Sky Park & Observatory***
7 ***Tomintoul and Glenlivet – Cairngorms National Park***
8 ***West Penwith***

Reserves

9 ***Brecon Beacons National Park***
10 ***Cranborne Chase***
11 ***Eryri National Park (Snowdonia)***
12 ***Exmoor National Park***
13 ***Kerry***
14 ***Moore's Reserve South Down National Park***
15 ***North York Moors National Park***
16 ***Yorkshire Dales National Park***

Sanctuaries

17 ***Ynys Enlli (Bardsey Island)***

Communities

18 ***The Island of Coll (Inner Hebrides)***
19 ***Moffat***
20 ***North Ronaldsay Dark Sky Island (Orkney Islands)***
21 ***The island of Sark (Channel Islands)***

Details of these sites and web links may be found at the IDA website: https://www.darksky.org/ Many of these sites have major observatories or other facilities available for public observing (often at specific dates or times).

Dark Sky Discovery Sites

In Britain there is also the ***Dark Sky Discovery*** organization. This gives recognition to smaller sites, again free from immediate light pollution, that are open to observing at any time. Some sites are used for specific, public observing sessions. A full listing of sites is at **https://www.darkskydiscovery.org.uk/** but specific events are publicized locally.

Glossary and Tables

aphelion	The point on an orbit that is farthest from the Sun.
apogee	The point on its orbit at which the Moon is farthest from the Earth.
appulse	The apparently close approach of two celestial objects; two planets, or a planet and star.
astronomical unit (AU)	The mean distance of the Earth from the Sun, 149,597,870 km.
celestial equator	The great circle on the celestial sphere that is in the same plane as the Earth's equator.
celestial sphere	The apparent sphere surrounding the Earth on which all celestial bodies (stars, planets, etc.) seem to be located.
conjunction	The point in time when two celestial objects have the same celestial longitude. In the case of the Sun and a planet, superior conjunction occurs when the planet lies on the far side of the Sun (as seen from Earth). For Mercury and Venus, inferior conjunction occurs when they pass between the Sun and the Earth.
direct motion	Motion from west to east on the sky.
ecliptic	The apparent path of the Sun across the sky throughout the year. Also: the plane of the Earth's orbit in space.
elongation	The point at which an inferior planet has the greatest angular distance from the Sun, as seen from Earth.
equinox	The two points during the year when night and day have equal duration. Also: the points on the sky at which the ecliptic intersects the celestial equator. The vernal (spring) equinox is of particular importance in astronomy.
gibbous	The stage in the sequence of phases at which the illumination of a body lies between half and full. In the case of the Moon, the term is applied to phases between First Quarter and Full, and between Full and Last Quarter.
inferior planet	Either of the planets Mercury or Venus, which have orbits inside that of the Earth.
magnitude	The brightness of a star, planet or other celestial body. It is a logarithmic scale, where larger numbers indicate fainter brightness. A difference of 5 in magnitude indicates a difference of 100 in actual brightness, thus a first-magnitude star is 100 times as bright as one of sixth magnitude.
meridian	The great circle passing through the North and South Poles of a body and the observer's position; or the corresponding great circle on the celestial sphere that passes through the North and South Celestial Poles and also through the observer's zenith.
nadir	The point on the celestial sphere directly beneath the observer's feet, opposite the zenith.
occultation	The disappearance of one celestial body behind another, such as when stars or planets are hidden behind the Moon.
opposition	The point on a superior planet's orbit at which it is directly opposite the Sun in the sky.
perigee	The point on its orbit at which the Moon is closest to the Earth.
perihelion	The point on an orbit that is closest to the Sun.
retrograde motion	Motion from east to west on the sky.
superior planet	A planet that has an orbit outside that of the Earth.
vernal equinox	The point at which the Sun, in its apparent motion along the ecliptic, crosses the celestial equator from south to north. Also known as the First Point of Aries.
zenith	The point directly above the observer's head.
zodiac	A band, streching 8° on either side of the ecliptic, within which the Moon and planets appear to move. It consists of twelve equal areas, originally named after the constellation that once lay within it.

The Constellations

There are 88 constellations covering the whole of the celestial sphere, but 24 of these in the southern hemisphere can never be seen (even in part) from the latitude of Britain and Ireland, so are omitted from this table. The names themselves are expressed in Latin, and the names of stars are frequently given by Greek letters followed by the genitive of the constellation name. The genitives and English names of the various constellations are included.

Name	Genitive	Abbr.	English name
Andromeda	Andromedae	And	Andromeda
Antlia	Antliae	Ant	Air Pump
Aquarius	Aquarii	Aqr	Water Bearer
Aquila	Aquilae	Aql	Eagle
Aries	Arietis	Ari	Ram
Auriga	Aurigae	Aur	Charioteer
Boötes	Boötis	Boo	Herdsman
Camelopardalis	Camelopardalis	Cam	Giraffe
Cancer	Cancri	Cnc	Crab
Canes Venatici	Canum Venaticorum	CVn	Hunting Dogs
Canis Major	Canis Majoris	CMa	Big Dog
Canis Minor	Canis Minoris	CMi	Little Dog
Capricornus	Capricorni	Cap	Sea Goat
Cassiopeia	Cassiopeiae	Cas	Cassiopeia
Centaurus	Centauri	Cen	Centaur
Cepheus	Cephei	Cep	Cepheus
Cetus	Ceti	Cet	Whale
Columba	Columbae	Col	Dove
Coma Berenices	Comae Berenices	Com	Berenice's Hair
Corona Australis	Coronae Australis	CrA	Southern Crown
Corona Borealis	Coronae Borealis	CrB	Northern Crown
Corvus	Corvi	Crv	Crow
Crater	Crateris	Crt	Cup
Cygnus	Cygni	Cyg	Swan
Delphinus	Delphini	Del	Dolphin
Draco	Draconis	Dra	Dragon
Equuleus	Equulei	Equ	Little Horse
Eridanus	Eridani	Eri	River Eridanus
Fornax	Fornacis	For	Furnace
Gemini	Geminorum	Gem	Twins
Hercules	Herculis	Her	Hercules
Hydra	Hydrae	Hya	Water Snake

Name	Genitive	Abbr.	English name
Lacerta	Lacertae	Lac	Lizard
Leo	Leonis	Leo	Lion
Leo Minor	Leonis Minoris	LMi	Little Lion
Lepus	Leporis	Lep	Hare
Libra	Librae	Lib	Scales
Lupus	Lupi	Lup	Wolf
Lynx	Lyncis	Lyn	Lynx
Lyra	Lyrae	Lyr	Lyre
Microscopium	Microscopii	Mic	Microscope
Monoceros	Monocerotis	Mon	Unicorn
Ophiuchus	Ophiuchi	Oph	Serpent Bearer
Orion	Orionis	Ori	Orion
Pegasus	Pegasi	Peg	Pegasus
Perseus	Persei	Per	Perseus
Pisces	Piscium	Psc	Fish
Piscis Austrinus	Piscis Austrini	PsA	Southern Fish
Puppis	Puppis	Pup	Stern
Pyxis	Pyxidis	Pyx	Compass
Sagitta	Sagittae	Sge	Arrow
Sagittarius	Sagittarii	Sgr	Archer
Scorpius	Scorpii	Sco	Scorpion
Sculptor	Sculptoris	Scl	Sculptor
Scutum	Scuti	Sct	Shield
Serpens	Serpentis	Ser	Serpent
Sextans	Sextantis	Sex	Sextant
Taurus	Tauri	Tau	Bull
Triangulum	Trianguli	Tri	Triangle
Ursa Major	Ursae Majoris	UMa	Great Bear
Ursa Minor	Ursae Minoris	UMi	Lesser Bear
Vela	Velorum	Vel	Sails
Virgo	Virginis	Vir	Virgin
Vulpecula	Vulpeculae	Vul	Fox

The Greek Alphabet

α	Alpha	ε	Epsilon	ι	Iota	ν	Nu	ρ	Rho	φ (ϕ)	Phi
β	Beta	ζ	Zeta	κ	Kappa	ξ	Xi	σ (ς)	Sigma	χ	Chi
γ	Gamma	η	Eta	λ	Lambda	ο	Omicron	τ	Tau	ψ	Psi
δ	Delta	θ (ϑ)	Theta	μ	Mu	π	Pi	υ	Upsilon	ω	Omega

Some common asterisms

Belt of Orion	δ, ε and ζ Orionis
Circlet	γ, θ, ι, λ and κ Piscium
Guards (or Guardians)	β and γ Ursae Minoris
Head of Cetus	α, γ, ξ^2, μ and λ Ceti
Head of Draco	β, γ, ξ and ν Draconis
Head of Hydra	δ, ε, ζ, η, ρ and σ Hydrae
Keystone	ε, ζ, η and π Herculis
Kids	ζ and η Aurigae
Little Dipper	β, γ, η, ζ, ε, δ and α Ursae Minoris
Lozenge	= Head of Draco
Milk Dipper	ζ, γ, σ, ϕ and λ Sagittarii
Plough or Big Dipper	α, β, γ, δ, ε, ζ and η Ursae Majoris
Pointers	α and β Ursae Majoris
Sickle	α, η, γ, ζ, μ and ε Leonis
Square of Pegasus	α, β and γ Pegasi with α Andromedae
Sword of Orion	θ and ι Orionis
Teapot	γ, ε, δ, λ, ϕ, σ, τ and ζ Sagittarii
Wain (or Charles' Wain)	= Plough
Water Jar	γ, η, κ and ζ Aquarii
Y of Aquarius	= Water Jar

Acknowledgements

Our thanks to Gerry Breslin and Samuel Fitzgerald at Collins; and to Ed Bloomer, Imo Bell, Catherine Muller and Sam Imperato, astronomers at Royal Observatory Greenwich, and Elizabeth Cunningham, Astronomer Emeritus at Royal Observatory Greenwich.

Image Credits:

pp. 16, 21, 35, 37, 41, 47, 53, 67 Shutterstock
p. 29 (left) NASA/Johns Hopkins University Applied Physics Laboratory/Carnegie Institution of Washington/Southwest Research Institute
p. 29 (right) Gerald Rhemann
p. 43 Artsiom Petrushenka/Alamy Stock Photo
pp. 49, 85, 97 A Fujii (ESA/Hubble)
p. 55 Taavi Niittee/Törva Astronoomiaklubi
p. 59 Vanderbei
p. 61 Stocktrek Images/Alamy Stock Photo
p. 65 W. Carter
p. 71 Davide De Martin (ESA/Hubble)
p. 73 REU program/NOIRLab/NSF/AURA
p. 77 Arnas Goldberg
p 79 Ljubinko Jovanovic
p. 83 Chris Guidry
p. 89 Misha Benshik
p. 91 Steve Jurvetson
p. 95 John Sanford/Science Photo Library
p. 101 Asim Patel
p. 103 Roberto Mura

Further Information

Books

Bone, Neil (1993), ***Observer's Handbook: Meteors***, George Philip, London & Sky Publ. Corp., Cambridge, Mass.
Chu, A (2012), ***The Cambridge Photographic Moon Atlas***, Cambridge University Press, Cambridge
Dunlop, Storm (1999), ***Wild Guide to the Night Sky***, HarperCollins, London
Dunlop, Storm (2012), ***Practical Astronomy***, 3rd edn, Philip's, London
Dunlop, Storm, Rükl, Antonin & Tirion, Wil (2005), ***Collins Atlas of the Night Sky***, HarperCollins, London
Ellyard, David & Tirion, Wil (2008), ***Southern Sky Guide***, 3rd edn, Cambridge University Press, Cambridge
Heifetz, Milton & Tirion, Wil (2017), ***A Walk through the Heavens***, 4th edn, Cambridge University Press, Cambridge
Heifetz, Milton & Tirion, Wil (2012), ***A Walk through the Southern Sky***, 3rd edn, Cambridge University Press, Cambridge
National Geographic & Wei-Haas, Maya (2023), ***Stargazer's Atlas: The Ultimate Guide to the Night Sky***, National Geographic, Washington, D.C.
O'Meara, Stephen J. (2008), ***Observing the Night Sky with Binoculars***, Cambridge University Press, Cambridge
Ridpath, Ian (2018), ***Star Tales***, 2nd edn, Lutterworth Press, Cambridge
Ridpath, Ian, ed. (2003), ***Oxford Dictionary of Astronomy***, 2nd edn, Oxford University Press, Oxford
Ridpath, Ian, ed. (2004), ***Norton's Star Atlas***, 20th edn, Pi Press, New York
Ridpath, Ian & Tirion, Wil (2004), ***Collins Gem – Stars***, HarperCollins, London
Ridpath, Ian & Tirion, Wil (2017), ***Collins Pocket Guide Stars and Planets***, 5th edn, HarperCollins, London
Ridpath, Ian & Tirion, Wil (2019), ***The Monthly Sky Guide***, 10th edn, Dover Publications, New York
Rükl, Antonín (1990), ***Hamlyn Atlas of the Moon***, Hamlyn, London & Astro Media Inc., Milwaukee
Rükl, Antonín (2004), ***Atlas of the Moon***, Sky Publishing Corp., Cambridge, Mass.
Scagell, Robin (2000), ***Philip's Stargazing with a Telescope***, George Philip, London
Sky & Telescope (2017), ***Astronomy 2018***, Australian Sky & Telescope, Quasar Publishing, Georges Hall, NSW
Stimac, Valerie (2019), ***Dark Skies: A Practical Guide to Astrotourism***, Lonely Planet, Franklin, TN
Tirion, Wil (2011), ***Cambridge Star Atlas***, 4th edn, Cambridge University Press, Cambridge
Topalovic, Radmila & Kerss, Tom (2016), ***Stargazing: Beginner's Guide to Astronomy***, HarperCollins, London

Journals

Astronomy, Astro Media Corp., 21027 Crossroads Circle, P.O. Box 1612, Waukesha, WI 53187-1612 USA. http://www.astronomy.com
Astronomy Now, Pole Star Publications, PO Box 175, Tonbridge, Kent TN10 4QX UK. http://astronomynow.com
Sky at Night Magazine, BBC publications, London. http://skyatnightmagazine.com
Sky & Telescope, Sky Publishing Corp., Cambridge, MA 02138-1200 USA. http://www.skyandtelescope.org/

Societies

British Astronomical Association, Burlington House, Piccadilly, London W1J 0DU. http://www.britastro.org/
The principal British organization for amateur astronomers (with some professional members), particularly for those interested in carrying out observational programmes. Its membership is, however, worldwide. It publishes fully refereed, scientific papers and other material in its well-regarded journal.

Federation of Astronomical Societies, Secretary: Ken Sheldon, Whitehaven, Maytree Road, Lower Moor, Pershore, Worcs. WR10 2NY. http://www.fedastro.org.uk/fas/
An organization that is able to provide contact information for local astronomical societies in the United Kingdom.

Royal Astronomical Society, Burlington House, Piccadilly, London W1J 0BQ. http://www.ras.org.uk/

The premier astronomical society, with membership primarily drawn from professionals and experienced amateurs. It has an exceptional library and is a designated centre for the retention of certain classes of astronomical data. Its publications are the standard medium for dissemination of astronomical research.

Society for Popular Astronomy, 36 Fairway, Keyworth, Nottingham NG12 5DU.

http://www.popastro.com/

A society for astronomical beginners of all ages, which concentrates on increasing members' understanding and enjoyment, but which does have some observational programmes. Its journal is entitled *Popular Astronomy*.

Software

Redshift, Redshift-Live. http://www.redshift-live.com/en/

Starry Night & Starry Night Pro, Sienna Software Inc., Toronto, Canada. http://www.starrynight.com

Stellarium, https://stellarium.org/

Internet sources

There are numerous sites with information about all aspects of astronomy, and all of those have numerous links. Although many amateur sites are excellent, treat any statements and data with caution. The sites listed below offer accurate information. Please note that the URLs may change. If so, use a good search engine, such as Google, to locate the information source.

Information

Astronomical data (inc. eclipses) HM Nautical Almanac Office: http://astro.ukho.gov.uk

Auroral information Michigan Tech: http://www.geo.mtu.edu/weather/aurora/

Comets JPL Solar System Dynamics: http://ssd.jpl.nasa.gov/

American Meteor Society: http://amsmeteors.org/

Deep-sky objects Saguaro Astronomy Club Database: http://www.virtualcolony.com/sac/

Eclipses NASA Eclipse Page: http://eclipse.gsfc.nasa.gov/eclipse.html

Ice in Space (Southern Hemisphere Online Astronomy Forum): http://www.iceinspace.com.au/index.php?home

Moon (inc. Atlas) Inconstant Moon: http://www.inconstantmoon.com/

Planets Planetary Fact Sheets: http://nssdc.gsfc.nasa.gov/planetary/planetfact.html

Satellites (inc. International Space Station)

Heavens Above: http://www.heavens-above.com/

Visual Satellite Observer: http://www.satobs.org/

Star Chart: http://www.skyandtelescope.com/observing/interactive-sky-watching-tools/interactive-sky-chart/

What's Visible

Skyhound: http://www.skyhound.com/sh/skyhound.html

Skyview Cafe: http://www.skyviewcafe.com

Institutes and Organizations

European Space Agency: http://www.esa.int/

International Dark-Sky Association: http://www.darksky.org/

Jet Propulsion Laboratory: http://www.jpl.nasa.gov/

Lunar and Planetary Institute: http://www.lpi.usra.edu/

National Aeronautics and Space Administration: http://www.hq.nasa.gov/

RASC Dark Sky: https://rasc.ca/

Solar Data Analysis Center: http://umbra.gsfc.nasa.gov/

Space Telescope Science Institute: http://www.stsci.edu/